John B. Armstrong, Texas Ranger and Pioneer Ranchman

Number Ten:
Canseco-Keck History Series
Jerry Thompson, General Editor

John B. Armstrong,

TEXAS RANGER AND PIONEER RANCHMAN

Chuck Parsons

Foreword by Tobin Armstrong
Afterword by Elmer Kelton

TEXAS A&M UNIVERSITY PRESS
College Station

Frontispiece: John B. Armstrong as a young man, probably in the mid-1870s.
Courtesy Tobin Armstrong

Library of Congress Cataloging-in-Publication Data

Parsons, Chuck.
John B. Armstrong : Texas Ranger and pioneer ranchman / Chuck
Parsons ; foreword by Tobin Armstrong ; afterword by Elmer Kelton.
—1st ed.
v. cm.—(Canseco-Keck history series ; no. 10)
Includes bibliographical references and index.
ISBN-13: 978-1-58544-553-0 (cloth : alk. paper)
1. Armstrong, John B., 1850–1913.   2. Pioneers—Texas—Biography.
3. Ranchers—Texas—Biography.   4. Frontier and pioneer life—Texas.
5. Ranch life—Texas—History.   6. Texas Rangers—Biography.
7. Outlaws—Texas—History—19th century.   8. Crime—Texas—History—19th
century.   9. Texas—History—1846–1950.   10. Texas—Biography.   I. Title.
II. Series.   F391.A73P37   2006   976.4'05092—dc22 [B]
2006011363

Armstrong [was] the big, handsome beau ideal
of what a Texas Ranger should look like. He also turned out to be
what a Texas Ranger should act like—he really loved a fight.

—DIANE SOLETHER SMITH, *The Armstrong Chronicle*

Contents

Illustrations

Foreword

The name Armstrong has a colorful history. Around the middle of the eleventh century a Scottish king had his horse killed from under him during a battle. His armor bearer, a powerful Dane, managed to lift the king onto another mount, thus saving his life. For such a heroic deed the warrior was granted a crest and the family appellation of "Strong Arm," or "Armstrong." Siward the Armstrong later became the Earl of Northumberland and, under the aegis of Edward the Confessor of England, led an army into Scotland to defeat the usurper, Macbeth.

Many generations later, and now subject to documentation, James and Jean Armstrong settled in the Shenandoah Valley of Virginia. One son, Martin Armstrong, served as a colonel in North Carolina in 1776. Various Armstrong men served as soldiers in America's early wars, and many became doctors, saving instead of taking lives.

Tobin and Anne L. Armstrong. *Courtesy Tobin Armstrong*

On May 23, 1842, Dr. John B. Armstrong Jr. married Maria Susannah Ready, the daughter of the founder of Readyville, Tennessee. Her father, Charles Ready, had fought as a volunteer under Tennessee's Gen. Andrew Jackson in the War of 1812.

This is only a very brief summary of the genealogical background of the subject of this book, Texas Ranger John Barkley Armstrong, but it portrays a little of his background, a background of highly efficient fighting men, intelligent ancestors, and leaders of the community. There was hardly a weak limb in the family tree.

John Barkley Armstrong certainly continued this honorable tradition. He chose to leave his native state at an early age and made his way west. Whether he was conscious of it or not is unknown, but he was following a much-honored tradition of moving ever westward. The family tradition of community service, although he certainly didn't think of it this way, was continued when he joined the Travis Rifles, a semimilitary group formed in the early 1870s to protect the City of Austin.

Following this service, which involved protecting a newly elected governor, he joined an elite group, McNelly's Special Force of Texas Rangers. He served valiantly under Capt. L. H. McNelly, then under Lieut. Lee Hall. Numerous times he was in charge of a detachment, having proven to his superiors that he was capable of leading men into dangerous situations and bringing them out alive after accomplishing his mission. In 1877 he was trusted with the assignment of capturing the most dangerous man in Texas, John Wesley Hardin, and with a single shot accomplished that mission. Other notable bad men also were brought in, thanks to the effectiveness of his leadership.

Following his Ranger service he proved to be a successful land dealer and rancher. He also became a Special Ranger without pay, not to justify his wearing a pistol or carrying a gun but to protect the property of such cattlemen as Richard King as well as his own land. He served as a commander of volunteers during the Spanish-American War, although he was unable to go to the battlefields himself.

This biography is the first full-length study of my grandfather, John Barkley Armstrong. We as a family are grateful for the research that has gone into this work, as much new information has been uncovered of which we were unaware. Writings which discuss Armstrong before this work are essentially about the bad men he pursued and captured, his role thus being minimized by the excitement provided by the lives of the outlaws. This biography is a deserving presentation of the life of a true American pioneer, John Barkley Armstrong.

— Tobin Armstrong

Preface

The subject of John Barkley Armstrong has been of deep interest to me for many years. In going through my correspondence I find that some letters are dated as early as 1969. At the start I doubt that I entertained seriously the thought of ever writing a full-length biography of the man, as in those early years my interest lay more in the collecting of information. But through the years ideas develop, or get dropped entirely, or maybe even forgotten. Perhaps it was a combination of factors that caused me to set other projects aside and concentrate on this particular Texas Ranger. Elmer Kelton's novel *Captain's Rangers* places Armstrong side by side with the great Capt. L. H. McNelly. My biography of McNelly was published in 2001, and perhaps the thought of focusing seriously on Armstrong came about at that time.

Through the years various of my ideas about Armstrong had to change as my research in primary source materials proved statements in popular writings wrong. In a sense, having to give up "established" ideas can be frustrating, but finding the truth of an incident is rewarding. A case in point is the "murder" of John Armstrong. This particular facet of the research is worth recounting.

J. Evetts Haley is considered a fine historian but was incorrect about Armstrong's passing. In *The XIT Ranch of Texas and the Early Days of the Llano Estacado* he writes how John Armstrong followed former Texas Ranger Ira Aten as foreman of the huge ranching operation. In Haley's work, Gene Ellison, a friend of Armstrong's, suspects J. W. Williams of stealing one of his cows, which leads to bad blood between the two men and with Armstrong. On November 18, 1908, at the Bovina depot in Parmer County, Williams shoots Armstrong off his horse, killing him. The first trial results in a hung jury; a second trial results in Williams's receiving a six-year sentence. The verdict is reversed, and at his third trial Williams receives a nine-year sentence.[1] Needless to say, for some time in my early days I accepted that the John Armstrong in Haley's book was John Barkley Armstrong. I was not alone.

Harold Preece believed as I did. In his biography of Ira Aten, *Lone Star Man: Ira Aten, Last of the Old Time Rangers,* he is not satisfied that Armstrong could be killed by a single man and adds to the myth: "Ira managed to maintain some precaution against rustlers but his successor, John Armstrong, was killed by a band of cattle thieves."[2]

Paul Trachtman further embellishes Haley. In the popular book *The Gun-fighters,* one volume in the Time-Life series on the Old West, he writes the following: "In the 1880s, when two big west Texas spreads, the LS and the XIT, sought to wipe out outlaws . . . ranch owners were able to attract top gunfighter talent. . . . One of the XIT gunmen was former Texas Ranger John Armstrong, the man who finally brought John Wesley Hardin to book."[3] Trachtman does not mention Armstrong's killing but is obviously following Haley and Preece.

Recently I reread the memoirs of Ranger Ira Aten, preserved in bound typescript at the Center for American History at the University of Texas at Austin. Aten speaks of Haley's having interviewed him in 1929, the year when *The XIT Ranch* was published. Aten, ten years a manager of the XIT and retired comfortably in California, is thinking back over the years after a visit from famed former Texas Ranger John R. Hughes. He recalls that when "cattle rustling was at a minimum on this great ranch I resigned as its manager turning the ranch over to John Armstrong. I instructed him before leaving how to handle the New Mexico cattle rustlers, that was to follow a policy of putting the fear of death into their hearts. Poor John, he was slack in his discipline in dealing with those men and they killed him about a year after I left."[4] Aten identifies the foreman as John Armstrong, not as John Barkley Armstrong.

Ira Aten's statement that John Armstrong was killed was embellished through the years by subsequent writers: Haley, Preece, Trachtman, and perhaps others. Maybe more people read the Time-Life book than the others, in which it is clearly stated that John Barkley Armstrong was the XIT gunman; earlier authors identify the man only as John Armstrong. There *was* a man named John Armstrong killed on the XIT Ranch in 1908. He had been division manager for eighteen years when he met his death by gunfire. This John Armstrong left a widow and one son. His brother Bill survived those difficult years by working on the Middlewater division of the XIT.[5]

As with any biography the author owes a great debt to others who contributed to the research. I owe a huge debt to Anne and Tobin Armstrong of the Armstrong Ranch, Armstrong, Texas. Mr. Armstrong was the grandson of the Texas Ranger and preserved a wealth of documents and photographs that added greatly to the biography. Others who have assisted with material or who have read portions of the manuscript include Donaly E. Brice of the Texas State Library and Archives,

Austin, Texas; Harold J. Weiss Jr., Leander, Texas; Frank Faulkner, director of Texana Division, San Antonio Public Library, San Antonio, Texas; Candice Du-Coin, Round Rock, Texas; Thomas C. Bicknell, Crystal Lake, Illinois; Lisa A. Neely, archivist of the King Ranch, Kingsville, Texas; staff members of the Texas State Library and Archives, the Austin History Center, and the Center for American History, Austin.

Without the able assistance of Texas A&M University Press's Mary Lenn Dixon, editor-in-chief; Diana Vance, acquisitions assistant; and Leanna Pate, marketing, the work would have remained only an idea. Thanks also to Kathy Bork, copy editor. My gratitude to them all.

Dedicated to the memory of Tobin Armstrong, son of Lucie Tobin Carr and Charles Mitchell Armstrong, grandson of Texas Ranger John B. Armstrong. Tobin was born July 9, 1923, and passed away on October 7, 2005. He was a constant source of inspiration throughout the preparation of this biography as well as preserver of the Armstrong papers and memorabilia.

And to the memory of my parents
William Thomas Parsons (October 24, 1901–November 16, 1984)
and
Vera L. Parker Parsons (September 14, 1899–February 7, 1993).
They continually inspired me to read and to appreciate history and the written word.

— CHUCK PARSONS

John B. Armstrong, Texas Ranger and Pioneer Ranchman

John B. Armstrong in the late 1870s. *Courtesy John M. Bennett*

Genesis of a Fighting Man

[Armstrong was] a dashing fellow, and always ready to lead a squad
of the Rangers on any scout that promised to end in a fight.

—PVT. N. A. JENNINGS

J OHN B. ARMSTRONG is best known as the Texas Ranger who captured and
brought back to Texas justice the notorious man-killer John Wesley Hardin.
The capture took place August 23, 1877, at the railroad station in Pensacola,
Florida. At the time Armstrong needed a cane to walk, having accidentally shot
himself, yet he handled his big Colt revolver well enough to kill one of Hardin's
associates and knock the man-killer unconscious long enough to assure his capture. No other arrest in Armstrong's impressive career equaled that of taking
Hardin. Ironically, Armstrong felt the arrest did not deserve the national attention it received.

Like so many Rangers who gained a degree of fame, Armstrong was not a native Texan. He was born in McMinnville, Tennessee, on January 1, 1850, the son
of Dr. John Barkley and Maria Susannah Ready Armstrong.[1] His name first appears in the written record in 1850, when the family was enumerated on the Cannon County, Tennessee, census. That year Dr. Armstrong claimed $4,000 worth
of real estate.[2] A decade later, now living in Warren County, Tennessee, the head
of household claimed $4,500 in real estate and $5,500 in personal property. At
that time the family, according to the census, consisted of Dr. and Mrs. Armstrong, sixteen-year-old Thomas and fifteen-year-old Mary; daughter Laura was
fourteen. The three younger children were ten-year-old John Barkley; "Levan-
der" (Lavanda), age seven; and "Batona" (Betavia), five years old.[3] All were Tennessee born.[4] Being the son of a successful doctor young Armstrong received a

better than average education. Yet in spite of the advantages of being a member of a successful family, young John B. chose to leave the state of his nativity.

It is unknown exactly why he chose to leave Tennessee to travel west. Many a young man in the southern states had difficulty adjusting to Reconstruction following the defeat of the South. His sympathies were certainly with the Confederates, who were returning home to find their homes destroyed and families broken up. Many got into trouble with federal troopers. Armstrong no doubt resented the presence of Union officials and those popularly known as "Carpetbaggers," but no record is extant accusing him of any actual problem; more than likely he chose to leave to avoid unpleasantness with the occupying forces.[5]

Once the decision was made to leave home it was perhaps only natural that he would have "G.T.T." (Gone to Texas), as many others did who found it necessary to leave their troubles behind and seek a new life. "Gone to Texas" was the explanation for many a young man's leaving home, going to a land which was half frontier and controlled by marauding Indians. Armstrong left Tennessee with a young friend, Charles Mitchell. Somewhere along the way the pair split up, but, according to family tradition, they renewed their friendship in Galveston during Christmas 1872.[6]

When and why Armstrong chose to settle in Austin is uncertain. It is known, however, that he was in the capital city at least by late 1872.[7] On June 7, 1873, a meeting was held "for the purpose of organizing an independent infantry military company of the citizens of Austin." At the first meeting of the new Travis Rifles M. D. Mather was elected captain and John S. Myrick, orderly sergeant.[8] By mid-June the company numbered forty-five members, and the quartermaster was ordered to requisition arms from the adjutant general. Uniforms were ordered later. Albert S. Roberts was soon elected first lieutenant, Dr. Edward Wise, second lieutenant. The Rev. E. B. Wright was chosen chaplain, while J. S. Myrick, David Walker, Andrew S. Donnan, Stephen T. Mitchell, and Nathan Meyer were named sergeants. Charles H. Webb, William B. Brush, J. M. Swisher, and T. M. Miller were named corporals.[9]

By August notices of the Rifles' meetings appeared regularly in the columns of the city's major newspaper, the *Daily Democratic Statesman*. Occasionally, the subject inspired light jocularity from the editor: "Yesterday we saw a cavalcade numbering about twenty passing down the Avenue just before dark. We first thought it might be a retreat from the Indians on the frontier, but afterwards learned it was the band of knights who are to immortalize themselves in real deeds of glory at the tournament about the last of this month. They are a brave and valorous set of knights."[10]

Exactly when John B. Armstrong joined the Travis Rifles is unknown, but by the end of 1873 he was a recognized member. On Christmas Day a tournament was held at the racetrack which was described as "a brilliant affair." Riding as "knights" were Armstrong, Henry B. Barnhart, W. B. Brush, Ed Gray, John P. Holmes, John F. House, S. T. Mitchell, William Mitchell, Smith Parr, J. G. W. Pierson, Horace Rowe, Frank Tamplet, Valentine O. Weed, and Charles C. Worthington.[11]

Just what these "knights" did in this tournament was not recorded, but perhaps it was a sort of special rodeo demonstration of horsemanship, with Henry Barnhart, the "most successful knight," galloping away with the grand prize, a silver cup. Horace Rowe was judged to be the most "graceful rider" and was awarded a meerschaum pipe. V. O. Weed had the smallest score and received the tin cup. "Everything passed off smoothly," reported the *Daily Democratic Statesman,* with "most splendid weather" and a "very large attendance." The day's activities ended with a "grand convention ball" described as "a magnificent affair" with "any number of magnificent ladies present." Henry Barnhart made a "pretty address" and then crowned Miss Clara Haralson the Queen of Beauty, and "everything passed off to the great pleasure and delight of all." [12]

Perhaps one of those "magnificent ladies" present was Mary Helena "Mollie" Durst, the daughter of the late James H. and Mary Josephine (Atwood) Durst, who would become Mrs. J. B. Armstrong within five years. At the time the widow Durst operated a "Select Private Boarding" house in Austin on the north side of Hickory (now 8th Street), between Brazos and San Jacinto streets.[13]

The idea of Comanches marauding close to Austin was of little concern to these nineteenth-century knights, but a real test came when the incumbent governor was to surrender the office to the newly elected governor—and refused. Reconstruction governor E. J. Davis was determined to remain in office, notwithstanding the results of the valid election. He called on the Travis Rifles to protect him. Instead, they took over the halls of government and guaranteed the inauguration of Richard Coke as the new governor. The *Daily Democratic Statesman* explained the situation best: just before the 3:00 PM session of the legislature began, Adj. Gen. Frank L. Britton placed the offices of the governor in the basement of the capitol under "a military guard" composed of "forty or fifty negroes and white Radicals." The appearance of this armed force "at once attracted the attention of everybody, and each and all, who had sufficient curiosity to see the position assumed, gathered about and looked at the novel scene with feelings of indignation and disgust." A group of young men "expressed their estimation of the proceeding with hisses and sneers," causing Travis County sheriff George B. Zimpelman, in an effort to prevent a disturbance, to order out a posse of armed men, composed principally of members of the Travis Rifles.[14] Zimpelman's force remained watchful the entire night, although nothing of note transpired, according to the *Daily Democratic Statesman.*[15]

Mary Helena "Mollie" Durst, the future Mrs. John B. Armstrong. *Courtesy Tobin Armstrong*

V. O. Weed, a private in the Rifles, remembered that they were called to the capitol by the pounding of a kettle drum beaten by a young boy. Weed, as Armstrong and the others almost certainly did, shouldered his gun and marched. Captain Mather explained that he had received orders from Adjutant General Britton to protect Governor Davis. On guard around the capitol building were Weed and a man named Shaw. Shaw was so nervous that he accidentally fired his rifle and shot off the brim of his hat. Weed complained to Sheriff Zimpelman about Shaw and was asked whom he wanted to be on guard with him. "John B. Armstrong," Weed replied. Armstrong was called up, and the pair stood guard the entire night without incident.[16] Although this was no doubt the first time John Barkley Armstrong was called to duty for the State of Texas, he had already firmly established a reputation of courage and dependability.

When Governor Coke was fully in control the *Daily Democratic Statesman* could treat the whole affair with lightheartedness. It reported that the Travis Rifles had "immortalized themselves in the manly efforts they made . . . to preserve the peace. The boys performed their duty like old soldiers, and have the thanks of the public." The paper also delighted in pointing out that the Rifles had been ordered out by one governor, "and before they were ready to report to duty another was inaugurated. Instead of serving Davis they served Governor Coke, in behalf of popular government in Texas." [17]

Whatever tension the men of the Travis Rifles may have experienced while guarding the new governor, it was all forgotten by the time of their "first grand military ball," held the evening of Friday, January 30. The *Daily Democratic Statesman* urged Austinites to turn out en masse, "thereby expressing their approval and gratification at the existence of the organization, and availing themselves at the same time of an opportunity of tripping the 'light fantastic' *en regle*." A competent committee had spared no effort to make the occurrence a success, the paper assured its readers. And of course it was a success, as any affair backed by the *Daily Democratic Statesman* naturally would be. Indeed, it was "largely attended and proved a brilliant affair." Even Governor Coke and Lt. Gov. Richard B. Hubbard put in an appearance and "paid their compliments to the company in a most pleasant and happy style." [18]

In mid-February Miss Mollie Durst was so impressed by the organization that she made public her desire to present an "appropriate and beautiful flag to the Travis Rifles," but only if she could get a sufficient number of Austin ladies to assist her. She was "admired by all who know her, for her loveliness, amiability of manners, accomplishments and talent," gushed the *Daily Democratic Statesman*. By now the leading newspaper of the state was describing the Rifles as "universally admired," its members as "respected for their noble bearing and moral worth." Editor John Cardwell assured his readers that Miss Durst's "generous proposal" would meet with "general approval" and garner the assistance "of the refined and intellectual of her sex in this city." [19]

Certainly to the surprise of no one, Miss Durst's generous proposal received the necessary approval of and assistance from the "refined and intellectual," and the flag was presented Friday evening, June 5, at 6:00 PM on the capitol grounds. A "large concourse" of ladies and gentlemen were on hand, but prior to the presentation Capt. M. D. Mather led "the gallant young men of the company" up Congress Avenue with "glistening bayonets and military tread," accompanied by a band and citizens of all ages. After some delay a passageway was cleared through the audience, and the Hill City Club marched up to the stand. Mollie Durst stepped forward and presented the flag "in a very clear tone of voice [with] appropriate and touching language." At the conclusion of her remarks, and "as the vibration of applause died away," the band played, fittingly, "Mollie Darling." Then Captain Mather accepted the flag on the part of the company. The whole affair "passed off in a very creditable and happy manner, and we trust that the ball which followed contributed much to the happiness of all the members of the company and the club." Mollie's speech was printed in full in the *Daily Democratic Statesman,* its most memorable line perhaps being, "This banner is emblazoned with the name of no sanguinary fields to perpetuate the memory of civil war. . . . Given by the ladies of Austin, it is proper that this flag should bear conspicuous on its folds the Lone Star of Texas, shining still with a luster as bright and untarnished as when it blazed in the front of battle before Mexican marauders. You have already given proof that in your hands the flag of the Lone Star State is safe." Mollie was followed by Captain Mather, who offered a few appropriate thank-yous and concluded by saying to the young ladies, "the hearts of the Rifles are at your command." The entire affair was followed by a military ball.[20]

Armstrong himself was impressing the citizens of Austin not only with his ability on horseback but also with his leadership ability. In June 1874 he received a petition requesting him to run for the office of city marshal on the Democratic ticket. This document was signed by L. H. Fitzhugh, J. P. Maloney, A. P. Wooldridge, M. D. Mather, O. Archer, L. E. Edwards, and "many others."[21] The petition, dated June 20 and addressed to John B. Armstrong, Esq., asked him to run for the office in the primary election on June 24. Armstrong, but twenty-four years old, exhibited his political sagacity by responding the same day that he did "highly appreciate the compliment" and pledged himself "to abide [by] the result of the primary election." And if nominated and elected, he would devote himself "with unflagging zeal to the discharge of its duties." In the primary Jeff Johnson received the greatest number of votes, 193. John B. Armstrong, one of eleven candidates, received only 72 votes. But when all the campaigning and counting was done, Ed Creary, running on the Independent ticket, was elected the city marshal of Austin.[22]

Although Armstrong impressed many an Austinite with his abilities, we have no record of what he did to so impress the politicians enough to cause them to ask him to run for such an important office. He never became city marshal. He remained in the Travis Rifles, no doubt for social reasons as much as anything else. What he did during this period to earn his livelihood is unknown.

On October 20 the Rifles held a target-shooting competition. In full parade uniform they marched from Congress Avenue to Graham's Place, a mile from the city. Company commissary colonel L. H. Fitzhugh provided a "most elegant lunch," which included "sparkling champagne that flowed from some twenty or thirty of Krug's best, and after two cakes had been done away with, dancing commenced in first class style." At one time there were sixteen couples on the floor, and after two waltzes a quadrille was called for. It was then "a sorrow for those that did not dance, not to be able to participate in the mirth and fun of the company."[23]

At 2:00 PM the target shooting began, two hundred yards from the bluff. Thirty-two Travis Rifles participated, among them, Armstrong, whose score was zero. Of the group, eighteen failed to score, perhaps because of the abundance of cake and champagne. Horace Rowe and Sam Wade both received gold medals for accuracy. Although eighteen Rifles had failed to score, the leather medal was awarded to a Mr. Gleason for being the poorest shot in the company. Three cheers were given for the winners and the captain, then the band played. After several dances the company marched home, "all merry and full of joy over the happy day they had spent. A good many ladies graced the occasion with their presence."

In November the Rifles were treated to a lavish supper by several young women, probably a Thanksgiving affair. The Committee of Arrangements printed a public thank-you in the *Daily Democratic Statesman,* naming Miss Phoebe Peck, Miss Annie DeCordova, Miss Clara Haralson, Miss Salome Smith, and—not surprisingly—Miss Mollie Durst for their generosity in providing the "magnificent supper."[24] The Rifles did "hereby unanimously return their sincere thanks for their exertions and kindness."[25]

Armstrong probably stayed in Austin because of Mollie Durst. Family tradition holds that the pair were attracted to each other early, but young John B. did not yet feel himself to be in a financial position to ask for her hand in marriage.

Austin was becoming too tame for Armstrong, a healthy, red-blooded young man yearning for action. Whereas the Travis Rifles offered him some degree of excitement and allowed him a place in Austin society, a new force had been created which would provide regular pay as well as the potential for a great deal of excitement and danger. In mid-1874 Governor Coke established the Frontier Battalion, a force of six military companies whose purpose was to protect the frontier. An adjunct force was also created to deal with more internal problems. A young Virginian, Leander H. McNelly, was chosen captain of this latter group. He was a veteran of four years of fighting for the Confederacy and for three years held the rank of captain in Gov. E. J. Davis's State Police Force. As McNelly was then residing in Washington County the special force was aptly named the Washington County Volunteer Militia Company. In 1875 Armstrong became a member of this company and thus a Texas Ranger.[26] His star would shine brightly in Ranger history.

Capt. Leander H. McNelly, circa 1872. From the original photograph taken in Montreal, Canada. *Courtesy Texas Ranger Hall of Fame and Museum, Waco, Texas*

CHAPTER TWO

Blood on the Palo Alto Prairie

Boys, I may lead you into hell; . . . I'll never send you into a battle,
I'll lead you. All I ask any man to do is follow me.

—CAPT. L. H. MCNELLY

DURING the latter part of 1874 McNelly and his command had been stationed in strife-ridden DeWitt County, where his orders were to quell the fighting between two groups, the followers of William E. Sutton and the extensive Taylor family. Only one Sutton was directly involved, but he had many friends and followers who were enemies of the Taylor clan, a large family whose menfolk were accused of horse and cattle stealing and other acts of desperadoism. One of their deadliest members was John Wesley Hardin, a first cousin of the Clements brothers, who were nearly as notorious as he was. McNelly kept squads of Rangers scouting throughout the country, dropping in unexpectedly at the "grog shops" to prevent serious plots to murder and waylay members of the other faction. McNelly's orders were to try and make the two groups become friendly, which proved to be an impossible task. His men were engaged in only one exchange of gunfire with members of the Sutton force. Had he stayed longer in that county perhaps he would have had to deal with more violent acts, but an event closer to the Rio Grande called for his special kind of leadership.

The records are conflicting as to when Armstrong actually was mustered into McNelly's militia. The July 31, 1875, muster and payroll prepared at Santa Maria in southwestern Cameron County records him as a private, mustered in on May 20, 1875. Another muster and payroll, dated August 31, 1875, and prepared at the "Magotee" (El Mogote) de Don Juan, shows Armstrong had been a private for three months and then was promoted to the rank of fifth sergeant on May 20, 1875. Likewise, the earliest service record available shows he "was mustered into the State Service" on May 20 with the rank of sergeant, although another document—dated May 9, 1875—records the value of the company's horses; Armstrong was riding a $130 horse. This document records the names of the men in the company, thirty-nine in number, at the Santa Gertrudis Ranch of Richard King. The men who determined the monetary value were King, Reuben Holbein, and William C. Chamberlain. The document was dated at Corpus Christi, Nueces County, and then recorded by the District Court clerk, Joseph FitzSimmons.[1]

Whichever date is accurate—May 9 or May 20—Armstrong was now a McNelly Ranger. For lieutenants McNelly selected T. C. "Pidge" Robinson, a fellow Virginian who not only handled the company's paperwork but also contributed lengthy and fascinating letters to the *Austin State Gazette* which kept Central Texas informed of what was happening on the border. James W. Guynn was second lieutenant. McNelly had need of five sergeants: George A. Hall; Roe P. Orrell; brothers Lawrence Baker and Linton Lafayette Wright, third and fourth sergeants, respectively; and Armstrong, now fifth sergeant. The earliest "reorganization" muster roll—dated August 31, 1875—shows McNelly had J. Brown as acting corporal and fifty-five privates making up his command. Listed also on the document are the names of fourteen men who were discharged, of whom five received a dishonorable discharge. McNelly, already suffering from advanced tuberculosis, the disease which would shorten the time he could lead his men in the field as well as shorten his life, may not have been physically strong but his orders had to be obeyed or else the Ranger was dismissed. Two men had also deserted by then.

One can only speculate about how long it would have taken McNelly to extinguish the fires of the Sutton-Taylor feud in DeWitt County had he been allowed to remain. However, a significant event happened down on the coast on Good Friday, March 26, 1875, which resulted in Governor Coke's sending McNelly and his troop there: the Nuecestown raid. Raiding and murder was common on both sides of the Rio Grande in the early 1870s, but reached a crisis point when a band of thirty or more raiders from Mexico ventured into Nueces County, intending to sack the city of Corpus Christi. The raiding party met resistance at Thomas Noakes's combination store and post office in Nuecestown, thirteen miles northwest of Corpus Christi. Noakes killed one of the raiders and a customer was shot down while leaving the store. Noakes managed to escape with his family, but the building was destroyed and all his property and the post office were lost. Area ranchers retaliated, forming mobs that they called "Minute Men" companies. They attacked the retreating raiders, losing one man but

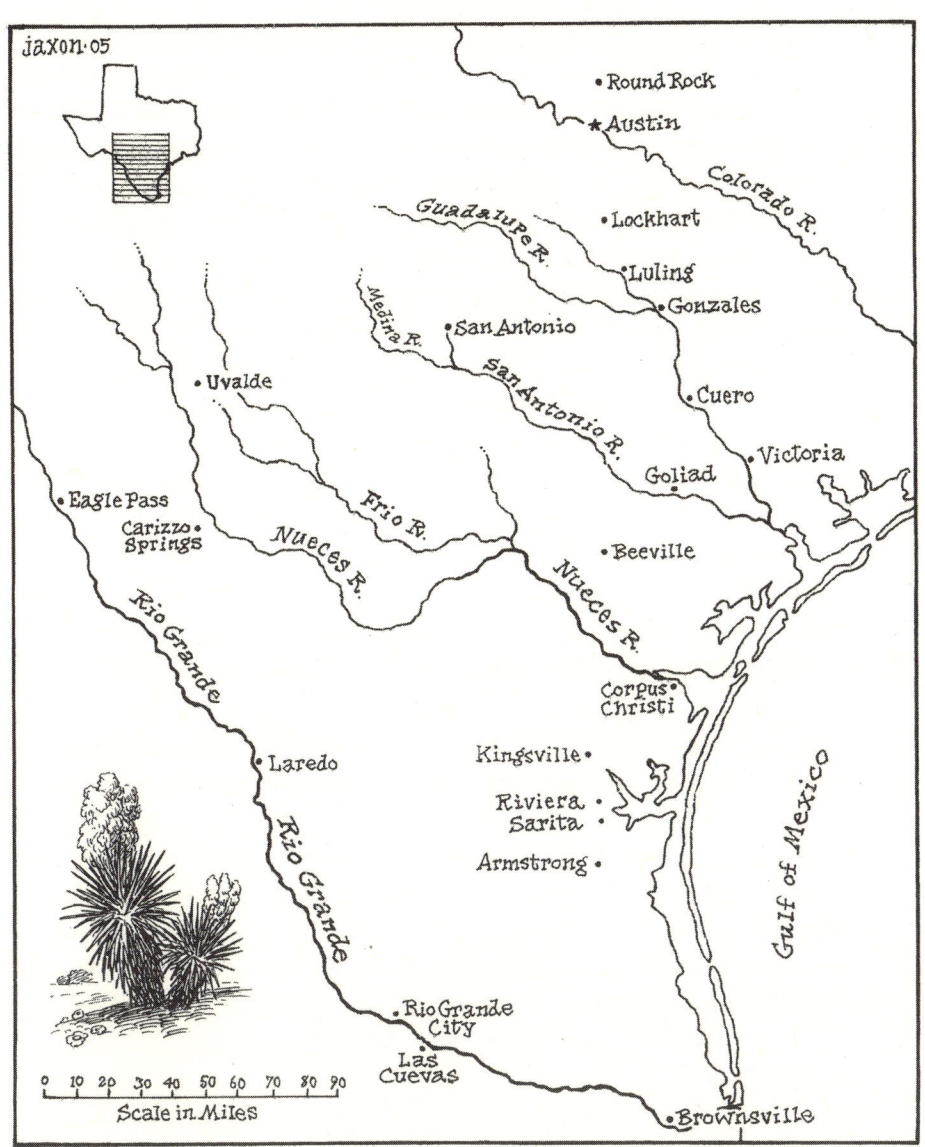

Map of South Texas, 1870–1915. *Drawn by Jack Jackson*

capturing a wounded raider, who was lynched. In the weeks following, numerous Mexican ranches were burned, and many Mexican citizens of Texas, probably innocent of any connection with the raiding party, were killed.[2]

Cattleman Richard King certainly requested help from Governor Coke, but more urgent was a plea from Nueces County sheriff, John McClane. On April 18 he telegraphed McNelly's superior, Adj. Gen. William Steele, in Austin: "Is

Capt McNelly coming[?] Five ranches burned last week by Mexican disguised."
Steele responded that McNelly's company had left ten days previously.[3] It was
into this dangerous Nueces Strip, that wild area between the Nueces River on
the north and the Rio Grande on the south, that McNelly would lead his men.
And his men would follow him without question.

A Ranger who served under McNelly in his later years recalled John B. Armstrong. Napoleon Augustus Jennings described him incorrectly as a Kentuckian,
but "like many of the sons of the Blue Grass State, was a giant in size." He recalled that Armstrong's "sweeping blond mustache and pointed 'goatee' would
alone have made him conspicuous among so many beardless young men," but
there was much more to Armstrong than his "hirsute charms." Jennings recalled
his "singularly mild blue eye" that his "handsome face was full of character." His
"carriage was as erect as that of a grenadier and, despite his great size, he was extremely graceful in all his movements." Further, Armstrong was "a dashing fellow, and always ready to lead a squad of the Rangers on any scout that promised
to end in a fight."[4]

Fellow Ranger George Durham, who served with McNelly longer than Jennings and who wrote a more objective account of his years in the Nueces Strip,
when meeting Armstrong for the first time noticed that he was "a big tall man
with whiskers on his chin."[5] A more objective description of Armstrong is con-
tained on the "Descriptive List" all Rangers had to carry in the 1880s. There he
is described as having brown hair and a light complexion, as thirty-eight years
old (in June 1888) and five feet eleven and one-half inches tall.[6] All descriptions
are correct.

One of McNelly's first actions upon arrival in the Nueces Strip was to disband
the various Minute Men companies acting out of a thirst for revenge against
raiders and suspected raiders. These groups were nothing more than organized
mobs of Anglo citizens who in all likelihood suspected any Mexican, no matter
on which side of the river he resided, to be a cattle thief and summarily executed
any they caught. McNelly was determined to be the law south of the Nueces and
would not tolerate any mobs moving unchallenged around the country. He had
the following order published in area newspapers to establish his authority: "In
consequence of the most recent outrages committed in this portion of the country by armed bands of men acting without authority of law, I find it necessary
to notify all such organizations that after the publication of this order I will arrest
all such bands and turn them over to the civil authorities of the counties where
they are arrested, and nothing but the actual presence of some duly accredited
officer of the county or State will protect them from arrest."[7]

In late April McNelly met with the leaders of several of these bands and convinced them that his authority was supreme and that there was no need for them to lead vengeance-seeking mobs across the countryside. Three of the Minute Men's commanders were Martin S. Culver, T. Hynes Clark, and Sidney G. Borden. Exactly what McNelly said to them has not been recorded, but the commanders disbanded their mobs without resistance.[8]

Although the young men of McNelly's command were anxious for adventure, that is, fighting bandits from across the river, McNelly was busy developing a spy system which would prove to be invaluable. The lack of action caused some recruits to consider quitting the force at this time. George P. Durham was one of these; in his later writings he does not identify what specifically caused him to remain in the service but he does recall a short speech the captain made causing him to disregard all previous thoughts about quitting and to keep McNelly's memory alive the remainder of his life. That little speech, which Armstrong heard as well and which perhaps proved to be inspiring when things got hot, was recorded by Durham as follows:

"Boys, I may lead you into hell; but I'll get you out if you do exactly as I tell you to do. I'll never send you into a battle, I'll lead you. All I ask any man to do is to follow me."

After that only death could prevent me from following McNelly.[9]

Between breaking up the unofficial Minute Men companies, which were no more than mobs in McNelly's eyes, Armstrong returned from San Antonio and Austin. We know nothing of his purpose in perhaps first going to Austin. He was one of eight passengers arriving in San Antonio on the stage from Luling, in Caldwell County, some forty-five miles southeast of Austin, who checked into the Menger Hotel.[10] Possibly, he was on business for McNelly—on "detached service"—which allowed him this time away from the company.

Whatever the reasons for Armstrong's absence from the Nueces Strip, on his return there was plenty of action, and death was to be close at hand in the first real action the Rangers saw there. While McNelly was developing his spy system he learned of raids on cattle and horses that were then driven across the river. In May three of his most dependable spies were in action: Herman S. Rock, of Brownsville; Lino Saldaña, a deputy sheriff of Cameron County; and Jesús Sandoval, described by Durham as "the most vicious, merciless killer that ever has come to my notice."[11] Their work bore fruit in the form of a decisive victory over the raiders in June on the old battle ground of the Mexican War, the Palo Alto Prairie. It was to be the most decisive victory over the raiders yet—and John B. Armstrong was there.

On June 5 McNelly learned that a party of raiders, at least sixteen in number, had crossed the river and were gathering Texas cattle to be crossed over. He immediately ordered Lt. T. C. "Pidge" Robinson and a squad of eighteen men to the crossing of the Arroyo Colorado, from where they were to send out scouts to learn when the raiders would return with their plunder. Three days later Sandoval had a prisoner, Rafael Salinas. With third-degree persuasion McNelly learned that there were sixteen raiders in the party, that it was under the command of Camilo Lerma and José María Olguín, alias El Aguja (The Needle). McNelly sent out another spy to learn the whereabouts of the raiders and to stay on their trail. Three days later another prisoner was brought in, Incosnascion [Incarnación?] García. After receiving the same brutal treatment as Salinas he told his story, which agreed with that of Salinas. He further confessed that the raiders had three hundred head of cattle stolen from Texas ranchers. McNelly hastily called his men together and headed in the direction of the Laguna Madre, arriving there about seven o'clock in the morning of the twelfth of June.

McNelly's orders were simple: his men were to form a skirmish line and advance upon the raiders; they were not to unsling their carbines or draw their pistols until McNelly fired the first shot; they were to pick out their target and stay with him, turning neither to the right nor the left unless the raider did. This would in theory prevent the Rangers from firing on one another. When the Rangers struck the water the raiders "commenced firing on us with Spencer and Winchester carbines." The Rangers, "not firing a shot or speaking a word, our line well dressed," advanced at a walk, since a more rapid gait on the wet ground was impossible. When the Rangers were within around seventy-five yards of their target, the raiders "wheeled their horses round" and rode off at a slow gallop. Soon the Rangers were on hard ground and were quickly within fifty yards, shooting distance. Even though now on firmer footing the Rangers could not overtake the retreating raiders. "So I ordered three of my best mounted men to pass in their right flank, and press them so as to force a stand, [and] as I anticipated, the Mexicans turned to drive my men off, but they held their ground, and I got up with four or five men, when the raiders broke." At that moment the raiders realized they were about to be overtaken and panicked. McNelly continues: "After that it was a succession of single hand fights for six miles before we got the last one. No one escaped out of the twelve that were driving the cattle. They were all killed." [12]

Actually, the number of raiders lost to Cortina was fifteen killed; one escaped. During the heat of battle one Ranger, L. B. "Sonny" Smith, was shot and killed. He was the youngest Ranger in the company. Rather than leave the bodies of the raiders for the buzzards, McNelly had Cameron County sheriff James Browne gather them up and take them to the Brownsville city square. There the corpses were piled up like stacks of cordwood, a challenge to their friends or relatives to come and take them. Ultimately, the dead raiders were buried in a trench.

Young Private Smith was given a state funeral compared with what the raiders received. His remains were escorted to the city cemetery by a large procession of citizens, soldiers from Fort Brown, as well as McNelly's men. Col. H. C. Merriam and the famous Col. John S. "Rip" Ford were "parade marshals." At 3:30 in the afternoon of June 13 Lieutenant Robinson, Armstrong, Sandoval, and the others marched to the Brownsville undertaker's, where a hearse was waiting, hitched to two big black horses. Two marching bands were also there to provide appropriate funeral dirges. Two companies of U.S. troops fired a salute over Smith's grave.

Later, McNelly learned just how deadly his attack on the raiding party had been: "I have just learned that we wounded two men in our fight that escaped, and one of them has since died. It seems there were fourteen in the party. As I only counted twelve as the firing commenced, and found twelve bodies after it was over, I supposed we had killed all." Later still, on September 30, McNelly corrected the numbers and wrote that fifteen raiders had been killed and one wounded, who had indeed escaped.

The Palo Alto Prairie battle was the first significant engagement against the organized raiders from south of the Rio Grande, and McNelly and his men had defeated them decisively.[13] But on the other side, the master raider himself, Juan Nepomuceno Cortina, could still plot additional raids. He had lost not only three hundred head of stolen cattle but sixteen of his *bravos*.

As newsworthy an event as the Palo Alto Prairie engagement was, McNelly would accomplish an even greater feat later that year. By crossing over into Mexico to recover stolen cattle he defied treaties between the two countries, killed a number of suspected cattle thieves, impressed Richard King by returning stolen cattle to his ranch—and this all without losing a single man. The act of crossing the river to pursue bandits was necessary, in McNelly's mind, because he had not been able to catch raiders on the north side of the river, since the Palo Alto engagement. McNelly wrote: "I have followed fifty herds of cattle to the bank of the Rio Grande, and I would see the stock on the opposite bank. The Mexicans dare me to cross the river and take them. They would say, 'Here are the cattle, come across and take them if you dare.'" The spy system obviously was not getting information quickly enough to McNelly for him to stage an effective coup.

McNelly would change all that come November.[14]

Juan Nepomuceno Cortina, ca. 1862–1865. Adj. Gen. William Steele considered Cortina "the recognized head and protector of all the cattle thieves and murderers, from Camargo to the mouth of the Rio Grande." *Courtesy Robert G. McCubbin Collection*

CHAPTER THREE

Gunfire at Las Cuevas

Orders were received to ride rapidly to Las Cuevas, alias Robber's Roost.
We obeyed them to the letter; we rode rapidly; fifty-five miles in six hours,
each man carrying one hundred rounds of ammunition.

—LT. T. C. "PIDGE" ROBINSON

McNELLY and his men had little to be proud of following the resounding victory over the bandits on the Palo Alto Prairie; there were no more significant victories against the raiders. Cortina's thieves were wary of these new *diablos tejanos* (Texan devils) from the north and rather than chance being run down and killed like the men on the Palo Alto Prairie they abandoned the stolen herds to escape with their lives.

What frustrated McNelly most was not that he couldn't locate bands of cattle thieves, Mexican or Anglo, but that the Rio Grande was an international boundary which law officers could not legally cross without expecting repercussions. His success on the Palo Alto Prairie would not be easy to duplicate; in fact, the thieves were in such awe, or fear, of McNelly that they would not cross the prairie at night.

McNelly continued to work on his spy system and managed to locate a number of the bandits who were willing to provide information—for a price. He was able to inquire into the character of the men who made up the bands of raiders and "selected those whom I knew to be tricky, and secured interviews with them." The result was that he was able to proposition a few to sell out their companions, promising to pay them more for their betrayal than they could make by continuing to raid. McNelly knew that in order to stop the raiding he had to know ahead of time when a raid would occur and when the raiders would cross the river with their stolen stock. McNelly noted that all of these potential traitors whom he approached readily entered into his plans and that without exception he found them to be "reliable and trustworthy." McNelly depended so much on these traitors that he agreed not to "interfere with their own individual stealing at all. I gave them liberty when I was not there in their neighborhood, to cross over with their friends."[1]

Not only did McNelly establish a successful spy system, he also managed to get one of his own men, Sgt. George A. Hall, infiltrated into Cortina's organization. McNelly testified before the U.S. House of Representatives that he "sent" Hall on board the vessel that was to deliver some five hundred or six hundred head of Texas cattle, "for the purpose of taking down the brands. He went in the character of a spy. . . . Cortina was present himself, with a force of probably 150 or 200 men, delivering these cattle."[2]

During the months following the Palo Alto victory McNelly worked on improving his spy system, actually getting some raiders to betray their comrades in an effort to smash the raiding rings. Results were not satisfactory, however, as too many times McNelly and his small band of Rangers managed to catch up with the raiding parties only after they had crossed the river. Rather than give up, he chose to solve the problem in typically dramatic McNelly fashion: ignoring international law, he followed the raiders into their own land. For McNelly, Sergeant Armstrong, and the other young Rangers, this would be their first venture into Mexico.

T. C. "Pidge" Robinson, now writing for the *Daily State Gazette* of Austin, some six weeks after the late November invasion, provides an account of the venture into Mexico in his humorous style: "I would have sent you an account of the invasion long, long ago, but was compelled by special orders to write out a true account of it, which has been awful straining to the mind; I am not accustomed to this, and have scarcely recovered from it yet; besides, I think I must have been sun struck in Mexico, or received some kind of shock; I have not felt well since; it may be the effect of the heavy dew on the river; but every time a shot is fired in my vicinity, a disagreeable chilly sensation starts with lightning speed from the back of my neck, and comes out at my boot heels."[3] Robinson was rarely serious in his contributions to Austin's newspapers, but he did provide a firsthand account of the invasion to recover stolen cattle.

McNelly gave considerable thought to invading a foreign country, even if "in hot pursuit." He met with Maj. A. J. Alexander from Fort Brown regarding the problem of raiders, and, according to McNelly, Alexander advised him in writing that he could "follow raiders anywhere." McNelly, eager to catch raiders and knowing his men wanted action as badly as he did, hoped to put Alexander "to the test in a few days." Further, writes McNelly, "I heard that the parties who buy most of the stolen cattel [*sic*] have contracted to deliver (18,000) eighteen thousand head" to Monterrey "within the next ninety days." Eighteen thousand head would mean an average of two hundred head of cattle stolen from ranches in Texas and delivered every day.[4]

On November 18 McNelly sent word to his Rangers to prepare to ride hard for the Rancho Las Cuevas, on the Mexican side of the river, as there was a herd about to be crossed. "Orders were received," writes Pidge, "to ride rapidly to Las Cuevas, alias Robber's Roost. We obeyed them to the letter; we rode rapidly; fifty-five miles in six hours, each man carrying one hundred rounds of ammunition."[5]

McNelly was not alone in trying to stop the thieves. Lt. Col. James F. Randlett from Fort Brown had learned from an unidentified "Mexican Citizen of Texas" that a party of fifteen thieves had crossed and would probably recross with stolen cattle soon. Randlett intended to intercept them while they were still on U.S. soil. Thus, at almost the same time, two military forces had the best of intentions to get to the river before the bandits could cross. Randlett with some thirty men actually did catch up with the raiders shortly after they crossed but did not pursue them into Mexico. He did fire on them, killing two and wounding another, at which time the thieves retreated further, leaving only a few head of cattle which had gotten stuck in the river's sand.

McNelly arrived and heard the shooting but, unlike Randlett, chose to cross over with his twenty-six Rangers. Although he did manage to commandeer a boat, it was an unworthy affair as it leaked badly. Writes Pidge of the crossing, "In single file and leading the horses, we gathered at the beautiful, the beautiful river and in silence commenced the embankment; on each side of the crossing crouched the men, gun [rifle] in hand, to protect the passage of the first boat; from below at another crossing echoed the sullen roar of the Springfields, where the regulars [under Randlett] were making a feint, while we went over undiscovered; with a very little assistance I could have made a feint myself about that time. After crossing two horses and getting them through the quicksands, it was calculated that with the best of luck and no accident all of them could be safely landed on the Mexican bank by Christmas Eve."[6]

Nevertheless, McNelly and his men did manage to cross over in the dilapidated vessel without incident. Pidge writes that the "tub" was in such poor condition that "it kept four men busy bailing to keep the nose of the man who paddled above the water. It only carried four; it might have carried more had there been room enough for them to bail, but there wasn't." Finally, all were crossed over, McNelly, guide Jesús Sandoval, and interpreter Tom Sullivan going over first, followed by Armstrong, Robinson, and George A. Hall, who got across with their horses.[7] The remainder were on foot. On the Mexican side they rested until dawn, then proceeded to march toward Las Cuevas, the rancho which Prv. William Callicott called the headquarters of all the thieves.[8] Pidge describes the exhausting march: "We . . . marched to Las Cuevas, said to be one mile distant; after we had walked about three, we concluded there must be a mistake somewhere, or that the town was marching too; a little further and we ran full upon the ranche."[9]

At the ranch which McNelly believed to be Las Cuevas, orders were given to ride in and shoot everyone except old men, women, and children. McNelly, Armstrong, Robinson, Sandoval, Sullivan, and Hall galloped through shooting and yelling, followed by those on foot. Pidge records that seven Mexicans were killed and nine wounded. But McNelly had attacked the wrong ranch. Sandoval had not been on the Mexican side for quite some time and instead of leading McNelly and the Rangers to Las Cuevas had led them to a ranch close by, Las Cucharas. When McNelly realized his mistake he could only gather his forces and anticipate the worst. He had not only killed possibly innocent people but he had lost the element of surprise. Juan Flores Salinas, titular head of the bandits around Las Cuevas, was close enough to have heard the shots and no doubt was preparing to attack the invaders. Pidge estimated that there were at least two hundred men ready to defend their country and their honor and protect the stolen herd as well. Callicott estimated the number at 250. Whatever the number, they vastly outnumbered McNelly and his twenty-six.

There was little cover now for McNelly and his Rangers so the only sensible thing to do was to retreat back to the river, where the bank would give them some protection. This they did, with Lieutenant Robinson leading the way and McNelly bringing up the rear. Pidge wrote later of the retreat: "I was awfully fatigued, but I scorned to let this interfere with my duty, and stepped up with much alacrity; such fiendish yells I never heard, but I could see very little; where the smoke came from the guns it hung like a pall, and I was not sorry to leave, for it had a very offensive odor to me. Back to the river we went, and waited further developments."[10]

The "further developments" were unexpected, as now, with Mexicans charging him, Captain Randlett came to the rescue with some forty soldiers of the 8th Cavalry. With all the gunfire Armstrong's and Hall's horses panicked and jumped out from under them, leaving them afoot. The two horses were taken by the Mexicans. McNelly now brazenly ordered his men to "open up." In the firefight which followed Juan Salinas was killed, resulting in the Mexican line's breaking and retreating. McNelly bent down and picked up Salinas's fancy pistol, a Smith & Wesson inlaid with silver and gold. The captain, having survived the first onslaught, was determined to stay in Mexico until he recovered at least some of the stolen cattle. No doubt, Armstrong and Hall, now on foot, were just as determined to stay until they recovered their horses.

And stay they did. McNelly crossed back to the Texas side of the river to communicate with President Grant about his actions in crossing and creating a potential political crisis. During his absence a group of Mexicans under a white flag and led by Dr. Alexander Manford Headley, an English doctor who practiced on both sides of the river, approached the troop with the intention of convincing the Rangers to return to Texas soil to save their lives. The doctor and his group were rebuffed.

Three times the truce party approached the Rangers, each time requesting the same thing: return to Texas. But each time the response was the same: they would leave only with the stolen cattle. Ultimately, McNelly did agree to return to Texas on condition that the stolen cattle and the horses and saddles of Armstrong and Hall be returned the next morning at Rio Grande City, a few miles up the river. McNelly almost certainly agreed to this because of Dr. Headley's involvement. It was Dr. Headley who negotiated the terms of peace which allowed McNelly to save face by returning to Texas and who promised that at least part of the stolen herd would be returned.[11] McNelly recorded this in a telegram dated November 21: "I withdrew my men last night upon the promise of the Mexican authorities to deliver the cattle to me at Rio Grande City this morning." [12]

But when the next morning came no cattle were produced, so again McNelly crossed, but this time he met with a delegation of citizens who informed him that the cattle could not be crossed because they had not been inspected. McNelly saw this as a delaying tactic and threatened that he "would Kill the last one of them" unless the cattle were produced within five minutes.[13]

Pidge recalled that the Mexicans needed assistance to cross the cattle, and McNelly and ten men, Armstrong one of them, went over.[14] Then another excuse was provided, that there was no permit allowing the cattle to be taken across the river. Pidge Robinson was one of the ten who helped with the crossing and recalled the following: "Capt. M[cNelly]—exhausted all arguments with these gentlemen, except one, which he reserved for the very last, and which, as a *dernier resorte* [*sic*] in this country, is considered 'a clincher;' then he exhausted that; 'Prepare to load with ball cartridges—load!' The ominous 'kerchak' of the carbine levers as the long, murderous looking cartridges were chambered home, satisfied them as to the permit and the cattle were allowed to cross over without one; such is the power of a fifty-calibre argument, such the authority of Sharp on International law." The cattle were miraculously produced under this threat, although the number amounted to only sixty-five head. McNelly telegraphed Adjutant General Steele that the Mexicans would produce "more [cattle as] soon as captured and the delivery of the thieves."[15]

The sixty-five head of cattle were delivered, and Armstrong and Hall recovered their horses, saddles, and bridles without further incident. No thieves were delivered, however, in spite of the promise of the Mexicans. The cattle were returned to their owners, many of them ranchers in the immediate area. Those of cattleman Richard King were also returned, with volunteer Rangers Durham, Callicott, Rudd, and Pitts herding them back to the Santa Gertrudis Ranch. A greatly surprised King had never expected to see them or any of the Rangers again, as he anticipated that McNelly and his invading force would become "another Alamo." He was so grateful that he ordered the right horn of each of the recovered cattle to be sawed off and the cattle turned loose on the range to live out their days in peace. King's vaqueros called these special cattle *los viejos* (the old ones).[16]

It is unfortunate that during the tumultuous affair no single incident resulted in Armstrong's receiving any special attention. He was one of the first to cross the Rio Grande and among the first to attack, although unintentionally, the Rancho Las Cucharas. Even though a sergeant, he, as well as Sergeant Hall and the privates, were merely following orders and had no opportunity to display heroics. He received some little attention when his horse was recovered. However, all this would change in 1876 and 1877. Glory enough would be his.

Espantosa Lake in 1997.

CHAPTER FOUR

Engagement at Espantosa Lake

A lively little fight ensued.

—JOHN B. ARMSTRONG

FOLLOWING the invasion of Mexico until late 1876, little is known of Armstrong's activities. He did reenlist for another quarter, as his name appears on the muster and payroll dated February 29, 1876, prepared at Laguna de las Flores by Lt. T. C. Robinson. This shows service was from December 1, 1875, to February 29, 1876, with the rank of fifth sergeant. Other sergeants were George A. Hall, Roe P. Orrell, and brothers Lawrence B. and Linton L. Wright.

On December 28 a scouting party under the command of an unidentified sergeant was on the Rincón de Perro ranch some forty miles north of Las Rucias, a mile north of the river in Cameron County, when they found a "perfect Golgotha of stolen hides," as Lieutenant Robinson described the grisly scene in his report. The sergeant, possibly Armstrong, but conceivably any of the others, arrested the ranchero nominally in charge, who then attempted to bribe them. When the sergeant refused to be bribed, the ranchero attempted to flee and was killed for his effort. No additional details are provided in Robinson's report.[1]

On May 17 McNelly and his men caught up with a party of four Mexicans crossing stolen cattle only five miles from Edinburgh (present-day Hidalgo).[2] Two of the raiders were killed, one wounded. Seven head of cattle were recovered as well as six horses and their equipment. This action occurred while Mexican general Mariano Escobedo was dining with U.S. officers at Edinburgh and within earshot of Escobedo's twenty-piece band.

McNelly had first requested Capt. Henry J. Farnsworth of the 8th Cavalry to assist, but he refused to cross the river. McNelly and three volunteers did cross and searched for but failed to find any sign of the stolen beeves. "Capt. McNelly then sent for the Alcalde of Reynsia [Reynosa], and demanded the return of the cattle and thieves. He promised both, but nothing was done. The impression prevails that the cattle were for Escobedo's troops."[3] If Armstrong was a member of that scouting party he would have undoubtedly volunteered to again cross into Mexico to recover stolen cattle. Surprisingly, the press gave very little attention to this second McNelly invasion, perhaps because so few animals were recovered, they apparently were not stolen from Richard King, and the "invaders" were in Mexico but a short time.

One significant event in which Armstrong no doubt participated, although again neither his name nor any other individual names appear in the available record, was the arrest of desperado John King Fisher in June 1876. By the time McNelly entered the Nueces Strip, Fisher was notorious, a thief and a killer, considered the leader of a large group of desperate men. Fisher, born in 1854, was four years younger than Armstrong and got in trouble at an early age. As a teenager he broke into a store in Goliad County and stole some trifling items. In October of the same year, 1870, he was arrested by Texas state policeman C. C. Simmons and charged with theft and robbery. Sentenced to two years in Huntsville State Penitentiary, he was pardoned after serving only four months. On April 6, 1876, he married Sarah Vivian, daughter of a prominent ranching family.

Although some credit Fisher with killing dozens of men, McNelly believed he had killed only nine. His reputation originated from the influence he held over the area around Pendencia Creek, a stream rising west of Carrizo Springs in northwestern Dimmit County and flowing into Zavala County. The lands which depended on that creek became known as "King Fisher's Territory." Fisher was considered the leader of many young men—some no doubt outlaws but some having joined him for protection from other desperadoes. He was intelligent, and his ability to manipulate the court system—and avoid prison—proved it. When he faced arrest by McNelly and his squad, he knew fighting was useless and surrendered, much to the chagrin of the young Rangers, who would have preferred a fight.[4]

Fisher's "home place" was not far from Carrizo Springs. McNelly divided his force into two groups and "rounded up" (surrounded) the house. Instead of fighting back, Fisher's men chose to surrender, no doubt on the orders of Fisher himself. Besides Fisher, McNelly arrested Burd Oberchain, alias Frank Porter; Warren Allen; William R. Templeton; Al Roberts; Bill Wainwright; Jim Honeycott; and Wes and Bill Bruton.[5] This arrest did lead to considerable press coverage of McNelly; the *San Antonio Daily Express* announced the arrest in bold headlines, followed by smaller headlines: "Captain McNelly Did It" and "His Men Scouring the Country for More of Them."[6]

Fisher was arrested on June 3, 1876, but two days later the group was released "whilst Captain McNelly was on his way with witnesses." Reportedly, seven of the nine could have been indicted for murder. They also had in their possession between six hundred and eight hundred head of stolen cattle and horses, which were turned loose because the brand inspector refused to inspect them.[7] McNelly resolved that in the future a fight was much preferable to bringing outlaws into court, as too many authorities were intimidated by them. Armstrong would face a similar situation later in the year with men supposedly of King Fisher's band, but the result was far different.

On July 25, in Victoria, Victoria County, the Washington County Volunteer Militia Company was mustered out and the next day reorganized as the Special State Troops.[8] Lt. Lawrence B. Wright was placed in charge of the arms and ammunition. Later, when McNelly was being replaced by Lt. J. L. Hall, before notary public Eugene Sibley, Wright swore that he had "at all times Kept a Strict watch over the arms of the State."[9]

A month later, on August 27, Armstrong was sent with a small squad to San Patricio, San Patricio County, to investigate the killing of former sheriff Edward R. Garner. Whatever they learned was not reported to Adjutant General Steele, as nothing is recorded in the monthly returns.[10] In September the company's base was at Oakville, Live Oak County. On the first the sheriffs of Bee and San Patricio counties requested assistance in calming a disturbance in Sharpsburg— then a small supply point a mile from the Nueces River but today a ghost town —in south central San Patricio County. Order was restored without any arrests. On the tenth the Live Oak County sheriff called for assistance, and a detachment was sent to Oakville. One Mexican was arrested and charged with murder and was placed in the Oakville jail. Armstrong was perhaps involved in all of these actions. What is certain is that he was the man in charge later that month —on September 22—when he was sent to Carrizo Springs with a detachment.[11]

On September 30 he was ready for a scout to round up a camp of outlaws believed to be part of the King Fisher gang on the shore of Espantosa Lake in north central Dimmit County.[12] Armstrong further believed that if King Fisher was there perhaps other important fugitives from justice would be with him, such as Alf Day and Frank and Tom Callison, all suspected of involvement in the robbery of the E. & H. Seeligson & Co. Bank of Goliad, Goliad County.[13] The lake had the reputation of being haunted, which caused many travelers to avoid the area, especially at night. Armstrong gathered up Rangers Thomas Netteville Devine, George Durham, N. A. Jennings, Thomas J. Evans, George W. Boyd, and A. L. Parrott, a formidable group who, with their Sharps rifles and Colt revolvers, and with Armstrong in the lead, experienced no fear whatever even though heading for the fearful Espantosa Lake.[14] Armstrong had several citizens with him to help identify any outlaws they might capture. He sent a detachment of ten men and "a number of citizens" under Corp. M. H. "Polly" Williams to the Pendencia to arrest a party reported there.[15] He then took the remainder of his squad and started for Espantosa Lake.

On the way they captured Noley Key, a young man who was reputed to be a horse thief and, if not an actual member of the King Fisher gang, a sympathizer. Key surrendered meekly, and Armstrong convinced him to lead them to the outlaws' camp. Just how this was accomplished is not revealed in Armstrong's report, but George Durham later recollected that Armstrong, or someone acting on his orders, put a noose around Key's neck and hoisted him a few feet off the ground until he agreed to lead the way. Even with Key's help the squad did not know the exact location of the camp but finally found a young man identified as Smith who agreed to lead them there. Key was taken along because there was no place to keep him secure.

When, near midnight, the group was ready to approach the suspected outlaw camp, Armstrong told Key that if he attempted to run he would be shot and then—in Key's presence—ordered the Rangers who were to guard him to shoot him if he tried to escape. Thomas N. Devine, one of the guards, left a detailed statement of the incident. He wrote that, when the squad arrived near the supposed camp and dismounted, Armstrong placed him and Thomas J. Evans in charge of Key and the horses. After some time he "heard one or two shots and then a general firing." He, Key, and Evans were all sitting on the ground, some ten paces from the horses. Several horses, not used to hearing the heavy firing of the loud Sharps rifles, became restless. He and Evans started to quiet them down but then "discovered Key in the act of running; and eight or ten paces from us. I called to him to halt, which caused Evans to turn round, then I think he also ordered him to halt." Key at first hesitated then increased his pace, "as tho' he had not been determined at first, but now made up his mind to try it." Evans and Devine repeated their order to halt and, as Key did not stop, they fired. He fell "and died in a few moments. He was trying to reach a thicket or wood distant, probably twenty or thirty yards from where we were." Key was some fifteen paces from them when they fired.[16]

Meanwhile, Armstrong and his squad, having left guide Smith and one of the citizens, a Mr. English, behind for their own protection, were advancing on the camp. With Espantosa Lake to their rear the outlaws would be trapped between the Rangers and the water. Apparently, it was Armstrong's intention to sneak into the camp and "get the drop" on them before the outlaws could resist. He assumed that they had a guard posted, as one of them, a hard case named Mullen, "was found the next day, in the lake with his clothes on and his cartridge belt around him." Armstrong believed this man was the guard and had fallen asleep, but when Armstrong called on the sleeping men to surrender, the sound awakened him. Instead of surrendering, the outlaws responded by firing their weapons; the Rangers had no alternative but to return fire. Armstrong reported the result: "The out-law replied by firing, when the troops opened fire killing Mullen . . . Roberts and Martin. Mullen, as before stated, getting into the lake. Martin had said around, that he would never surrender." The dead outlaws were all found to be undressed and in their blankets.[17]

When Armstrong and his squad returned to their horses they learned of the shooting of Key. They then headed for Thompson's ranch, where they hoped to find Alf Day and the Callison brothers. They were not found, but Armstrong arrested two suspicious characters. He had to release them the next morning because they could not be positively identified. Armstrong sent Smith to Carrizo Springs for a wagon to bring in the wounded McCallister and, presumably, Key and the dead trio killed on the shores of Espantosa Lake. Armstrong left an account of the bloody fray, reporting to Steele that they "discovered the camp on the bank directly in front of us" and then advanced slowly to within twenty yards, "when two commenced firing on us with their six-shooters." The Rangers "responded promptly, and a lively little fight ensued," with the killing of three outlaws and the "wounding of another in five places." Armstrong, perhaps with the assistance of guide Smith or Mr. English, identified the dead as John Martin, Jim Roberts, George Mullen, and Noley Key, "all of whom were desperate characters and the terror of this county." The wounded man was Jim McCallister, who had "but lately joined the party." Armstrong gave him water and made him as comfortable as possible on his blanket.[18]

Corporal Williams's squad also saw some shooting. "Learning that there was

Historical marker at Espantosa Lake, 1997.

a bad Mexican at Whaley's ranche, eight miles distant," wrote Armstrong, Williams "sent three men to arrest him. He refused to surrender, and fought desperately until our men were obliged to kill him in self-defense."[19]

Armstrong concluded his report to Captain McNelly by explaining that the parties they had killed on Espantosa Lake had about fifteen head of stolen horses and twenty-two yoke of oxen in Thompson's pasture, stolen from Mexicans on the Rio Grande.[20] The *San Antonio Daily Express* editor commented that Sergeant Armstrong's "blow" would "do more good toward restoring peace to our outraged frontier than any previous work of anyone." It was the first "effectual

effort since the re-organization of McNelly's company," and McNelly had informed him that he would keep on "pressing the bad men troubling our border until they disappear by the influence of the vigilantes and the bravery and
activity of the gentlemanly men under his command." The editor hoped that
McNelly (and, by extension, Armstrong) would "meet with the fullest success"
in their good work and "sweep clean our fair section of the dastardly thieves
who prowl about its recesses seeking what they may steal and whom they may
destroy." [21]

McNelly's October monthly return again shows the effectiveness of Armstrong and his ability to bring in wanted men. The return simply identifies three
men as being arrested on October 15: Charles Fox, wanted for attempted murder in Refugio County and taken to Atascosa County and jailed; Robert Rice,
brought in for theft in Refugio County and delivered to the Goliad County jail;
and William Davis, charged with assault to murder in Bee County and delivered
to Bee County sheriff D. A. T. Walton. [22]

Apparently, there had been some disrespectful words directed toward the
Rangers, at least in the Atascosa County area. When the *San Antonio Daily Express* reported on Armstrong's bringing in the prisoners, it quoted a Pleasanton
citizen's comments on the appearance of the Rangers: "them men don't look or
act like ruffians." The editor commented, "We did not hear them answer or see
them drink or act like soldiers often do, but behaved themselves in a quiet and
gentlemanly manner. Such acts are commendable and will be of great value to
themselves and the community in which they work." [23]

Armstrong's successful work at Espantosa Lake resulted in considerable favorable press coverage. The *San Antonio Daily Express* editor met with the Rangers on October 27 and commented in the newspaper on how Armstrong had "so
gallantly, with a portion of McNelly's rangers, attacked, killed and dispersed a
band of robbers not long since, belonging to King Fisher's marauders. The state
is to be congratulated in having in her service such a gallant body of men." [24]

On October 30 there were several arrests noted on the monthly return that
attracted no newspaper coverage, possibly because Armstrong was not involved.
Quillen Miller, alias Jake Jones, charged in Coryell County with liberating prisoners from the McLennan County jail and cattle theft, was arrested in Goliad
County. Also, Frank Callison, a Goliad bank robbery suspect, was arrested that
same day in Goliad County. Both Miller and Callison were delivered to San Antonio's notorious "Bat Cave" jail. [25]

As the end of 1876 approached, it became apparent that McNelly's health was
deteriorating rapidly. Dr. E. Melon of Brownsville, one of his numerous doctors,

sent a report to Adjutant General Steele indicating that he had treated McNelly during fifteen visits from April 1 to May 24 for fever, tapeworm, and pulmonary disease.[26]

Two days before, on November 13, McNelly was strong enough to send in a report dated November 15 to Steele. During the week ending November 11 he had sent one scout to Bandera County to serve a capias placed in his hands by County Attorney Anderson. Another scout was sent to Guadalupe County to serve papers for the court of Bexar County. Still another scout—made up of Pvts. T. W. Deggs and T. M. Quesenberry—was sent out to investigate the robbery of Jackson's store in Harwood, Gonzales County. Four men—identified as Frank Wingate, G. W. Keith, Ed Wingate, and Howard Little—were arrested in Wilson County on November 7 for this robbery. Two of the four were taken to storekeeper Jackson, who identified them as the culprits. They were delivered to the jail in Gonzales, Gonzales County. John Day and Jack Ryan of DeWitt County were also arrested during this same scout. "A very bad case of thirty or forty indictments against each of them, for theft & other serious offenses," reported McNelly. Another scout of three men was sent to Blanco, Gillespie, and Brown counties with subpoenas. "Have information on part of a band of horse thieves that opperate [sic] from here west—being located somewhere in Gillespie Co. Some maybe followed to Brown Co." A squad of eleven men under Jesse L. "Red" Hall was sent to Eagle Pass, Maverick County, the next day.[27]

Among the many scouts that were sent out over the South Texas landscape was one comprising Sergeant Armstrong and Pvt. T. W. Deggs, with orders to arrest one John Lewis Mayfield. Mayfield had been found guilty of the murder of Robert Montgomery in Parker County during the February term of the District Court in 1873. He was sentenced to death by hanging but, while waiting for the result of his appeal to the Texas Supreme Court, escaped from jail and headed south. Gov. Richard Coke offered a reward of $500 for his arrest and delivery "inside the jail door" of Parker County.[28] How McNelly learned of Mayfield's presence in Wilson County is unknown, but the fugitive was reportedly living in Graytown in the western portion of the county, some eleven miles from Pleasanton. Armstrong and Deggs were ordered to investigate and left San Antonio the evening of December 5 to arrest "a notorious character by the name of J. L. Mayfield."[29] The pair went to Mayfield's house, where they found him breaking in a colt.[30] Armstrong informed him of who they were but he responded with an oath, drew his pistol, and fired. Armstrong tells it best: "On the morning of the 7th instant Mr. Deggs & myself went to the house of one John Mayfield in Wilson County to arrest him for the murder of Robt. Montgomery in

Parker County—he resisted, fired twice at us & we killed him."[31] The *San Antonio Daily Express* was more dramatic, reporting that Armstrong "let him have a dose of cold lead, Deggs followed in the defensive, and Mayfield was in a few minutes a corpse." Apparently, the only person who witnessed the shooting was Mayfield's wife, who was inside the house. Deggs and Armstrong returned to San Antonio, "procured assistance, and went back to have an inquest held upon the body of the deceased."[32] The Rangers had believed they would be able to deliver Mayfield alive, but when they were fired on, it was a simple matter for them —kill or be killed.

The rival of the *San Antonio Daily Express,* the *San Antonio Daily Herald,* also reported the killing, adding that, when told he was under arrest, Mayfield "said he would be damned if he was, and drew his pistol and fired three shots before he was killed. He received two wounds, either of which would have caused his death, which was almost instantaneous."[33]

Although the inquest has not survived, a *San Antonio Daily Express* reporter apparently did have access to it. The verdict, strangely enough, was that Mayfield "had been murdered by one Armstrong and another man, both claiming to belong to McNelly's Rangers." On the twelfth the reporter obtained a copy and reviewed the evidence on which the verdict was based, probably the statement of Mayfield's wife, who may or may not have witnessed the killing. Commented the editor of the *San Antonio Daily Express,* "That jury must be a rare set. The evidence they base their judgment upon was that of a woman, who was inside of the house when the shooting occur[r]ed, and who was not close enough to hear what passed between the parties. What good citizens they must be, and what kind hearts they must have, that even a murderer, hiding from the law, and resisting arrest by the officers of the law, call[s] forth their strongest sympathies." The editor, obviously disgusted with the verdict, continues in this vein, recommending that the "good citizens" should "emigrate to the South Sea Islands and offer themselves as subjects to appease the appetites of the man-eating inhabitants of the lower regions."[34]

N. A. Jennings wrote that friends of Mayfield gathered immediately after the shooting, "men coming from every direction with their guns. [Armstrong and Deggs] knew that Mayfield had many friends, and they consequently thought discretion the better part of valor, jumped into their buggy, and drove quietly away, arriving at our camp two hours later."[35] Jennings was correct. Lt. L. B. Wright was made aware that the "prevailing sentiment . . . was not very favorable to this Company" and gathered Armstrong and Deggs and ten others to investigate. Wright talked to the sheriff, the county attorney, the district clerk, and "some of the best citizens [and learned] that the affair was heartily indorsed [*sic*] by the good people of the County." [36] Undoubtedly, Armstrong and Deggs regretted having to flee from Mayfield's friends, but there was no purpose in fighting them, as they knew they had been in the right and these Graytown citizens were acting out of anger rather than reason. Lesser men might have battled back, killing some, but Armstrong chose to avoid a possible conflict. Neither he nor Deggs was a coward.

Collecting the reward for Mayfield's apprehension proved to be impossible. Two days after the killing Armstrong wrote to Steele pointing out that he and Deggs had "seen a reward of five hundred dollars offered by the Governor for his arrest. Can we get it now since he has been killed?" [37] Steele responded six days later, pointing out that the reward offer stipulated that, since Mayfield had not been delivered inside the jail door of Parker County, the reward would not be forthcoming from the state.[38] During Armstrong and Deggs's absence Mayfield's friends had buried his corpse in a secret place, thus preventing them from producing the fugitive's body.[39]

Although Governor Coke failed to reward Armstrong and Deggs for their dangerous work, perhaps they did receive some tangible remuneration. J. N. Montgomery, a son of the deceased Robert Montgomery, later gave Armstrong one of his best horses. He also had offered a reward of $500 for Mayfield's arrest or killing, but whether this was paid is not known. Armstrong did travel to Parker County, where he met with "Dock" Lindsey, a lodge brother of Montgomery's, who provided a letter of introduction for Armstrong. The letter, written at Weatherford, the seat of Parker County, reads in part: "Dear sir & Bro this will be handed you by Mr Armstrong who . . . is the man that attempted to arrest John Mayfield. . . . He has been at some expence [*sic*] in hunting and finding Mayfield[.] Do for him in the way of remunerating him That you can[.] I think he is entitled to something." [40]

Armstrong may have received $500 from the Montgomery family, or possibly his only reward was the fine horse. If he did receive the $500, he no doubt shared it with Private Deggs. The Lindsey letter suggests that Armstrong might have ferreted out Mayfield's whereabouts.

These two Rangers were not the only men of McNelly's Company who had to shoot to kill. Cpl. M. H. Williams, who had been on the scout with Armstrong the night of September 30–October 1, 1876, in Dimmit County, was in charge of a squad delivering Jack Wingate, wanted for murder, to San Antonio. Williams had three privates with him when on the night of December 20 they were fired on from ambush by unknown parties. The horse of Pvt. Horace Rowe, former Travis Rifleman with Armstrong, was killed. Prisoner Wingate started to make his escape but, when ordered to stop, refused and was shot and killed. The inquest jury determined that McNelly's men were justified in their action. McNelly's monthly return reports the incident prosaically: "Killed in attg to escape while an attempt was made to rescue him was being made." [41]

Two nights later, Lieutenant Hall, with sixteen men, made one of the most spectacular arrests in the history of the Special State Troops. Hall and a squad arrested seven men charged with the murder of Dr. Phillip H. Brassell and his son George T. Brassell without firing a shot. [42] The double murder was committed by disguised men in DeWitt County the night of September 19. Other family members were able to identify several of the murderers, and, ultimately, warrants were issued. Why the Brassells were killed has never been satisfactorily determined. Possibly, the mob intended to do away with George alone, but when the doctor recognized them he had to be killed as well. It might have been part of the long-running Sutton-Taylor feud, or it might have been a killing having nothing to do with that trouble.

The opportunity to arrest all of the suspected killers came as a result of the marriage of Joe Sitterle and Melinda O. Cox on December 20. Only a month earlier William D. Meador had married Amanda Augustine, the daughter of David Augustine. On the night of December 22 the wedding dance was to be held at Clinton, formerly the DeWitt County seat, and it was Hall's belief that all of the indicted men would be present. His theory proved to be correct, and Hall and his squad did arrest the seven indicted men: Joe Sitterle, Jake Ryan, William Meador, Jim Hester, William Cox (brother of Sitterle's bride-to-be), David Augustine, and Charles Heissig. No list of the sixteen men with Hall has been preserved, so we do not know whether Armstrong was present. What is significant is that, although McNelly was still in command, he was so weak he could not have been able to travel to Clinton to make the arrest. [43]

Hall and Armstrong were both very capable men, and it would not be long before Adjutant General Steele would have to decide who was to become the new commander of the Special State Troops. Would it be Sgt. John B. Armstrong or Lt. Jesse L. Hall?

Capt. Jesse Leigh "Red" Hall, 1873, a few years before taking command of the Special State Troops. *Scribner's Magazine* 5, no. 9 (July 1873)

CHAPTER FIVE

John King Fisher Again

We have been riding hard ever since we left Eagle Pass
& our horses are very much jaded.

—JOHN B. ARMSTRONG

O N JANUARY 20, 1877, the Special State Troops were reorganized. McNelly had made them one of the top fighting units in Texas history but their continuation depended on the amount of funding the legislature provided. Furthermore, the end of the pay period gave Adj. Gen. William Steele the perfect opportunity to place a new man in command, as McNelly's medical bills were too high to retain him. And with McNelly unable to be in the field regularly the decision was not difficult. He had to be dismissed and a new man put in his place, a young and healthy man who had proven his ability to accomplish a mission and who could lead. Just how long Steele pondered the matter or whom he may have conferred with in making his decision is unknown, but he chose Jesse L. "Red" Hall to be first lieutenant commanding. Hall, according to Steele, "had already been represented by the people amongst whom he had been serving, . . . as the right man in the right place, and who was in the full vigor of early manhood and health." [1] The fact that Hall had also been sergeant at arms in the state senate may have provided the political motivation for Steele to choose him over Armstrong, who also was "in the full vigor of early manhood and health."

Sgt. John B. Armstrong was now promoted to second lieutenant. Some believed he should have been named the new commander; after all, he had been with McNelly on the Palo Alto Prairie fighting raiders in June 1875, and had been part of the invading force in November 1875. With his proven leadership abilities with the Travis Rifles and with greater experience than Hall (who had been in the service only since August 10, 1876), a good argument could have been made for him to succeed McNelly. George Durham, who had seen Armstrong in action, perhaps says it best. "As a man," he wrote, "Lieutenant Hall was all right. He had been city marshal of Denton [Sherman] and had been sergeant at arms of the Texas senate. He knew his way around. But most of us Rangers felt that if Captain McNelly was being fired either Sergeant Armstrong or Lieutenant Robinson should be promoted. So we didn't fall over ourselves welcoming Lieutenant Hall."[2] But the decision had been made. As far as is known Armstrong never expressed his feelings to anyone over the matter.

Hall put everyone to work immediately. On January 31 he and Armstrong and seven men went on a scout on the Nueces and Frio rivers, going as far as the edge of Atascosa County. Here they split up, Armstrong taking four men and going toward Pleasanton, then the county seat, while Hall took the remainder to Dog Town in McMullen County, where they arrested two men charged with stealing cattle and selling the hides.[3]

Armstrong, on February 2, arrested John Parker, wanted in Goliad County for a murder committed over three years earlier. He was delivered to the Atascosa County authorities. On the same day he arrested Ed Gillette, charged with theft of a cow. He too was carried to Pleasanton but was able to give bond. Armstrong, having learned of a cattle-skinning operation, continued with his scouts. The result was disappointing: "On scout Armstrong and party went to arrest some Mexicans skinning on West Bank of Nueces, but they had left as they were notified of coming of troops by Jack Hanning. Also went after B[uford] Cline — not found."[4]

Meanwhile, Sgt. O. S. Watson was working up an important case in Karnes County to arrest "the criminal" Alfred Allee on February 7.[5] Hall intended to head for San Patricio County on the ninth "to look into some cattle stealing in that section." At this point he felt communication was becoming more and more difficult with Adjutant General Steele's office. "I have only two men with me here," he wrote, "the others being scattered in different directions. The largest number in any one place being at Clinton. I have instructed the Sergts. to report to you occasionally as I cannot tell where they are on a long scout what they are doing and I will be off so much myself from headquarters that I cannot make a correct report unless they were with me now." From this initial report to Steele it is apparent that Hall was taking his work seriously and conscientiously. "It cripples me very much," he continues, "to have to keep so many men at Clinton but it is absolutely necessary to keep seven or eight men there for a while at least." Clinton, of course, was the focal point of the Sutton-Taylor feud, which had been raging for almost a decade. McNelly and his men had been there from July 1874 to early 1875, and had prevented any serious difficulties due to their constant patrolling, but McNelly had been sent to the Nueces Strip, leaving Clinton and DeWitt counties with only ineffectual local law enforcement. Although sheriffs from many counties were requesting Ranger presence, Hall felt his men were most needed in Eagle Pass while District Court was in session. "The people," he maintained, wanted Rangers present during their District Court, "and say if we will come out and back up the Juries and witnesses that we can indict King Fisher & his gang for murder in a number of cases and make them stick." Hall felt that if there was any way his troop could be increased to thirty men, instead of the mere twenty he had, it would "greatly increase the efficiency of my force."[6] This would allow him to have eight or ten men at Clinton and have twenty more for active service elsewhere.

Meanwhile, Armstrong and his squad were working in Atascosa and surrounding counties. On February 19 they arrested William Creswell in Frio County, and later they nabbed Fred Eatswinger in La Salle County. Frank M. Drake was arrested in McMullen County the next day, and J. H. Drake and F. M. Franklin were also arrested in Atascosa County on the twentieth. All were turned over to Atascosa County authorities.

A trace of frustration is evident in Armstrong's report, as he states how they had "rounded up" numerous houses and cow camps without finding the parties they were searching for. Sometimes a scout would stand guard around a place all night and find nobody there in the morning except women and children.

Covering these counties was especially arduous for the horses. "We have been riding hard ever since we left Eagle Pass & our horses are very much jaded," explains Armstrong in his report. In fact, one of "the boys" had to walk into town the day before as his horse was too weak to carry him. Armstrong estimated that his men averaged thirty to thirty-five miles every day on horseback. But soon they would be leaving for Goliad to get fresh mounts. Fortunately, the "good citizens" of the county had promised to get ten good horses and put them in a certain pasture near town and keep them there specifically for the benefit of the company. "This will be a great help to us," he concludes.[7]

Returning from Eagle Pass, Armstrong informed his superior that "large numbers of stolen cattle together with the thieves are now located at some point on Devils River & the head of the Llano." The cattle, stolen primarily from ranchers on the Nueces, were being held until ready to be driven in the spring either north or in the direction of New Mexico Territory. Because of these cattle thieves and numerous other fugitives, Hall felt is advisable to "make a raid on them as soon as the grass rises or as early as possible."

While in Eagle Pass, Armstrong searched for the Goliad bank robbers, as Hall had heard they were hiding out there. There were rumors they intended to rob a store in San Diego, Duval County, but they changed their plans, apparently having learned of the Rangers' presence, and fled the area.[8] The desperadoes were fleeing from the Rangers elsewhere as well. A correspondent for the *San Antonio Daily Herald* reported that in Dimmit County small bunches of horses had been run off by thieves from Mexico, and, because of the proximity of the Rio Grande, there was little use in pursuing "these ubiquitous kleptomaniacs, so we must grin and bear it." Besides the problem of horse theft, there were thieves who had "an extensive business in skinning cattle and getting away with the hides" to the southwest. But the situation was not beyond help, as a "timely visit of Sergeant Armstrong with four troopers caused a stampede for *al otra lado* [the other side]."[9]

Armstrong and the other Rangers still may have experienced a certain degree of anxiety over the killing of Noley Key and the others that night on the shore of Espantosa Lake. Key's mother and one of Mullen's brothers had attempted to bring a charge of murder against them. What must have been transmitted to Armstrong, and possibly to McNelly as well, was the contents of a letter from the county attorney of Maverick County, Theodore Terry, that there would be no further cause to worry. An inquest had been held on the bodies of Key and the others, the result forwarded to the grand jury, but no bills had been found against the Rangers. "The sentiment of all good men in this country is in favor of what was done at the time, & really it is *the only corrective*. We want more of it, and no jury can be impanelled [*sic*] that would convict McNally's [*sic*] men, for we have suffered too much here, and our county is so sparsely settled, that the due administration of law is very difficult." In fact, Terry had never heard anyone in his county complain about Armstrong or his command "except the immediate relatives of those killed, or parties who sympathized with them for the reason they are in the same business. *Let us have more of the same kind of business.*"[10] If Armstrong, Devine, T. J. Evans, or any of the others had ever given serious thought to the possibility of being indicted for murder, this letter laid their fears to rest.

More and more, the press gave attention to Armstrong and the Rangers who rode with him. He and Hall and Sgt. William L. Rudd were in Goliad the week ending on Saturday, March 16. Their presence there was reported in the *Galveston Daily News* as if the entire state would consider this important to know.[11]

Although the various newspapers of South Texas praised Hall and Armstrong for their good work, there are some incidents about which virtually nothing is known. One example is the arrest of Wallace Carter. The monthly return fails to mention Armstrong's arrest of this individual, but the *Dallas Daily Herald* carried a one-line item: "Sergeant Armstrong, of Captain Hall's command, arrested Mr. Wallace Carter on a warrant from East Texas." [12] All we know now is that the arrest was made in Wilson County. It must have been a peaceful arrest, as even the monthly return makes no mention of it.

In April Armstrong was in charge of a seven-man detachment stationed at Clinton, DeWitt County, as one never knew when the embers of the Sutton-Taylor feud would flare up again. Further, there was a detachment of six men guarding the jail at Beeville, Bee County, and always two or three men at Oakville, Live Oak County. Hall was in Austin at the time, working on a case involving a stagecoach robbery, and there were two men sick and unavailable for scouting duty. "I have remained in camp at this place, expecting every day the arrival of Lieut. Hall. Thought best not to make a move lest it should interfere with some of his plans," wrote Armstrong. [13]

Armstrong now had suspected bank robber Frank Callison to guard. On April 4 he sent Pvt. T. W. Deggs and three men to San Antonio to take Callison from that place for trial in Goliad. Once in Goliad they were to guard the jail throughout his trial. Then on the fifth Armstrong sent Cpl. A. L. Parrott with two men to Wilson County, where they succeeded in arresting Scott Raider, charged with horse theft. He was delivered to the Bee County sheriff. Since murderer Frank Singleton's execution was scheduled to take place on April 27, an additional three men were sent to make sure the execution was conducted without incident. With Corporal Parrott in charge there were no disturbances. [14]

On April 21, a Saturday, Armstrong, Sgt. O. S. Watson, and Pvt. S. N. Hardy were in Sutherland Springs in northern Wilson County, some twenty miles east of San Antonio. Why they were there is uncertain. The editor of the newly founded local newspaper, the *Western Chronicle*, E. R. Tarver, "had the pleasure of forming their acquaintance and found them elegant gentlemen. As soon as night had set in the detachment moved out of town, for what point is only known to themselves. We apprehend, however, that before the week is out we will be called upon to chronicle the arrest of some of the law breakers. We are glad to know that this command is sustained and upheld by all good citizens, and well does [it] deserve the confidence bestowed." [15]

Some scouts produced no results except tired horses and men. On April 27 Armstrong left Clinton with a detachment of five men and met with Sheriff

Reed of Kansas in Wilson County. The plan was to arrest one Willis Jackson for whom Reed had a warrant, but the lawmen failed to locate him. The detachment then left for Bexar County with the intention of arresting a Dr. Brazell, for whom Armstrong had capiases from several counties and other parties, but did not succeed in making any arrests. While in San Antonio, on April 30, they met with Atascosa County sheriff George W. M. Duck and "his efficient Deputy Mr. Young." They had learned of a party of eight or twelve outlaws in Atascosa County and they would have no difficulty in rounding them up. Young guided the group to where they were suspected of hiding. Armstrong "was surprised as well as disappointed" to find the place empty. The outlaws had been there the day before and had left, so the Rangers started to trail them. On May 3 they had to give up the chase as the trail was lost. The scout was not a total waste as they did make two arrests: on April 30 Charley Lewis was arrested, charged with theft of a cow in Llano County; and on May 2 Samuel Kirkandall was arrested, charged with theft of a mule in Bell County. The pair were turned over to Deputy Young and the Bell County sheriff was notified.[16]

Atascosa County was "quiet with the exception of the depredations committed by the band who infest a section some 20 miles north of this place commonly known as 'Sand Hollow.' They have stolen a great many horses & have recently threatened the life of some good citizens." Armstrong was not totally optimistic about his efforts and, in fact, might have been very disappointed: "We have been remarkably unfortunate in making arrests on this scout, notwithstanding the fact that we have worked hard. Bad characters are getting wilder every day & it is seldom that one ever sleeps in a house." Armstrong concludes his report by stating that he will leave for Eagle Pass the next day. On the road his hard work did pay off, as he made two arrests. He arrived in Eagle Pass on May 12 and made several additional arrests.[17]

Lieutenant Hall had plans for a "roundup" of outlaws in the Eagle Pass area, similar to what Maj. John B. Jones of the Frontier Battalion had done in Kimble County in April 1877, when he entered the county and arrested every man who could not give a good accounting of himself. Hall, Armstrong, and Sgt. Parrott were to arrive together and make the roundup. Hall arrived on May 11 and found Armstrong and Parrott already there, Armstrong having arrived on the ninth and Parrott on the tenth. A "general stampede of criminals had taken place to the Mexican side," but several characters were taken before they could cross the river.[18]

Hall had been delayed in getting to Eagle Pass. After leaving Clinton with a squad of seven men, he went through Goliad to Beeville in order to witness the execution of Frank Singleton. Parrott with his eight-man detachment was there guarding the jail. Singleton was executed on the twenty-seventh, and all passed quietly. Hall went to Oakville, then to Pleasanton, but was delayed there several days due to the ordnance wagon's breaking down. When it was repaired he continued with Deputy Sheriff Young to make some arrests but succeeded in making only two, although they took in ten or twelve "suspicious characters whom we had to release after several days of confinement not knowing where they belonged." In Medina County no arrests were made, nor in Uvalde County, mainly because the sheriffs were absent. At Fort Clark, District Court was in session, and Hall found that the "civil officers seem to be doing their duty in that County." Hall comments further that "the principal troubles it seems is from border thieves, who are very numerous," but Lt. Pat Dolan of Company F of the Frontier Battalion was doing "efficient service." Hall continues: "We have everywhere been kindly received and treated by the people, who have shown each detachment of my men every kindness and courtesy in their power, and specially were we warmly welcomed by the law abiding people of this County, who looked upon our arrival as truly a godsend just at this time, as in no section of the State have the good people and the law been worse trampled upon by desperadoes and thieves than in this County & Dimmit which is attached to this Co. for judicial purposes."[19]

Especially good news for Hall was that in Dimmit County members of the grand jury were no longer intimidated by outlaws. Grand jury members were notified that they would be amply protected by the Rangers. The outlaws had long held sway in the county and were accustomed to intimidating potential witnesses against them, but that was due to change. "I am happy to state that upon these assurances and our presence they seem to be doing their full duty as they have found bills against King Fisher and others who have gone scots [sic] free heretofore. [These outlaws] have been guilty of the most heinous crimes known to the law, committed even in the presence of the officers of the law," Hall wrote to Steele. District Judge Thomas Paschal and District Attorney John C. Sullivan did all in their power to assist him. The attorneys were not only "of learning and ability but fearless in the discharge of their duty." They were especially helpful in extraditing prisoners from Mexico.[20]

Upon Hall's arrival he found that Armstrong already had four prisoners under arrest on the Texas side of the river and had induced authorities on the Mexican side to arrest three men who had fled at the approach of the Rangers. Hall obtained requisition papers and immediately went with Armstrong and Parrott to Piedras Negras, on the opposite side of the river from Eagle Pass. The Rangers found and arrested a man known as Williams on the street and after "due formalities brought him a prisoner to this side."[21] Williams was an alias of Jim Jones, who, with Joe Horner and a man known as Murray, had robbed the Eagle Pass–Castroville stagecoach on April 17 in Uvalde County.

Hall, Armstrong, and the "gallant company of State Rangers" passed through Uvalde on May 21. A telegram was promptly sent to the *San Antonio Daily Express* informing that newspaper of the group's progress.[22] On Wednesday citizens of Castroville, in Medina County, saw the group as King Fisher was placed in the local jail. "King Fisher is here without his staff," quipped the newspaper editor. "He is well pleased but makes no public promenades."[23]

San Antonio's *Daily Express* gave considerable coverage to their action in the Eagle Pass area. "Lieut. Hall and His Men" was one headline in boldface, followed by a second in slightly smaller boldface print: "They arrive with a Superb Gang of Scoundrels." The story tells of how in recent weeks the Rangers had "captured thirty-three notorious thieves and desperadoes" and were "distributing them to the places where their presence is desired." King Fisher, with six indictments against him for cattle rustling and one for murder, and Andres Porter, also indicted for murder, had been left at Castroville. Jim Jones, alias Williams, also indicted for horse stealing in Maverick County, was to be taken to Austin. John Murray, charged with "the most diabolical rape on record" and who had broken out of jail at Pleasanton, would be kept in San Antonio. William Bruton, with fifteen indictments against him for horse stealing, and Bob Lewis, "a notorious horse thief," were both delivered to the Goliad jail. E. Kirkpatrick, also identified as J. B. Kirkpatrick, alias Dorn, a horse thief arrested at King Fisher's ranch, was sent to San Patricio County. Besides theft he was charged with forgery. Others included Charles H. Thayer, accused of embezzlement in Grimes County, who was sent to Anderson County; James Burdett, accused of being an accomplice of Ben Thompson's in the December 1876 killing of gambler Mark Wilson in Austin, was delivered to the Austin jail; J. F. West, an accused horse thief, was delivered to the Mason County jail. The *Daily Express* pointed out that in the "zealous discharge of his duties" Lieutenant Hall and his men had crossed into Mexico and secured the arrest of a number of the above-named men, who were "ignorant of extradition treaties" and who "felt themselves secure" on that side. "Too much praise cannot be awarded to officers like Lieutenants Hall and Armstrong," lauds the newspaper, "who have accomplished so much for the public good at such great risk to themselves. Criminals are fast finding out that 'the way of the transgressor is hard' in Texas, and are disposed to migrate to the Black Hills."[24]

Hall had reason to be proud. From the time he had left Clinton they had made thirty-three arrests but had had to release ten or twelve because they didn't know who they were or where they belonged, even though they knew that they were fugitives from somewhere. They had fled from the direction of Kimble County, where Major Jones had been operating successfully. These men invariably were going under assumed names, so it was difficult to know where they did in fact belong. Hall had some twenty-one men under arrest "with a certainty of a number more of arrests before we leave this place and before we reach Clinton." [25]

Hall expected the case against King Fisher to result in a change of venue. He also had to be concerned with appropriations to support his men. He intended to send Private Deggs to the stockmen, who had made propositions as to how much money could be raised and on what terms. "I will be extremely sorry if some arrangements cannot be made to retain my company in service. It is now in every way more efficient and composed of a better class of men than ever before, besides I fear the whole country where they have operated will relapse into the same condition as before the organization of the company especially on this border, particularly at this point & below." [26] Hall believed there was a great need for the force around Laredo, perhaps for several months.

Hall concludes his lengthy report by indicating that he might be going on an expedition after Indians with Col. W. R. Shafter of the U.S. Army. He states that if he does go on the raid he will leave Armstrong with a detachment "to wind up the court." [27]

Hall anticipated being in Clinton by June 1. Armstrong would then go to Austin for the DeWitt County prisoners being held there for safekeeping because of the inadequacy of the county jail. "If there is a small force in DeWitt during court there is likely to be trouble but with as many as 25 men there will be no danger." [28]

John Wesley Hardin ("Swain"), Florida, 1875. *Courtesy Robert G. McCubbin Collection*

CHAPTER SIX

Facing the Man Killer

Arrested John Wesley Hardin . . .
Had some lively shooting.

—J. B. ARMSTRONG

A LTHOUGH Hall knew he had good men under his command, men capable of tracking and arresting dangerous fugitives as well as handling themselves capably in a gunfight, he also realized that an accidental gunshot might take them out of action at any time. This is what happened to not only one of his men but two—Sgt. Oliver S. Watson and John B. Armstrong.

The details are scant. The first official report comes in the form of a telegram dated May 29, 1877. Hall informed Adj. Gen. William Steele that Armstrong had been accidentally shot that morning, "very seriously wounded in the hip I think," but not fatally. He also informed his superior that Watson had been accidentally shot through the neck twelve days before, on May 17, at Yorktown in DeWitt County. Although both men were temporarily out of service, Watson was "doing well." [1]

Armstrong was in the Case Hotel in Goliad when the accident happened. The *Victoria Advocate* describes him as "one of Capt. Hall's most trusty and bravest officers" who, while handling his own six-shooter, let it discharge, the ball entering the pit of his stomach and supposedly lodging in his thigh bone. The bullet was not yet extracted. It was considered a "dangerous and possible mortal wound," and he was in a "very critical condition in his room at the hotel." He was receiving the best medical attention possible. Two weeks later Armstrong was "improving and thought to be able to be up in a few weeks. Wound healing rapidly, ball not yet extracted." [2]

On May 31 Hall took time to write his superior a more detailed account of the two shootings. "I telegraphed you of Lieutenant Armstrong's misfortune," Hall, who may have discussed the wound with the doctor, begins, "who accidently [*sic*] shot himself in Goliad while carelessly handling his pistol. The ball entered just below the groin, striking the hip bone near the joint and up to this time the doctors have been unable to ascertain the location of the ball, but think it is lodged back of the hip bone. He is very severely wounded but I think the chances are in favor of his recovery. There is a probability however that he may be lame for life from the shot." Editor E. R. Tarver of the *Western Chronicle,* who had only recently made the acquaintance of Armstrong and was very favorably impressed, shared the bad news with his readers, writing that the "citizens will regret to learn that the gallant Lieut. Armstrong, . . . was dangerously wounded by the accidental discharge of his own pistol. . . . The ball entered the pit of his stomach and lodged against the thigh bone. Up to last accounts the surgeon had been unable to extract it. Though his condition is represented as critical, yet we hope this brave officer will soon recover, and be at his post again." [3]

Hall clearly told Adjutant General Steele of Watson's accident as well. Sergeant Watson was "accidentally wounded" by a pistol falling from a counter in a store in Yorktown, the ball entering his neck. "The wounding of these two officers," writes Hall, was "very unfortunate as I have no man in the company who can fill their places." He reminds Steele in this letter that both had "at all times" been "efficient and untiring in the discharge of their duty." [4]

No medical examiner's report has been found, so the exact nature of Armstrong's wound remains a mystery. It was serious enough to keep him out of action for some time, not only preventing him from participating in scouts but also keeping him from normal mobility. Once on his feet he had to use a cane. By early August he was able to "limp along without his crutches." [5] In an ironic turn of events the accidental self-inflicted gunshot wound was most fortuitous: being out of the saddle Armstrong could concentrate on going after the state's most wanted fugitive—John Wesley Hardin—who had been out of Texas since 1874.

From his teens until mid-1874 Hardin had traveled the state at will, ranging from the Corpus Christi area north to Hill County, or east to Louisiana; in 1871 he traveled with a cattle herd north to Abilene, Kansas, during which time he killed several individuals. Upon his return, always ready and willing to kill a real or imagined enemy, he joined the Taylor forces in their feud against the Suttons. On May 26, 1874, in Comanche, Comanche County, northwest of Austin, he killed Brown County deputy Charles Webb. In the following few weeks, Rangers under Capt. John R. Waller and various mobs lynched or shot to death eight of Hardin's relatives and friends, including his older brother, Joseph G. Hardin. Webb's killing resulted in a significant reward offered by the state for Hardin's arrest. On July 3, 1874, of that year Governor Coke increased the reward to $1,800 "for the arrest and delivery of the said John Wesley Hardin to the Sheriff of Comanche County inside the Jail door of said County."[6] Hardin feared mob action and with his wife and children fled Texas and settled in Florida, then in Alabama, where his wife—the former Jane Bowen—had relatives. For some time authorities had no idea where he was, this young fugitive who, by January 1877, was "worth" $4,000 to anyone who could deliver him to jail.[7]

In spite of rumors concerning his whereabouts and random sightings, state authorities did not actually know where Hardin was. Before he could be captured someone had to find where the deadly killer was hiding out.[8] Lewis Nordyke suggests that Hardin's hiding place was discovered through an intercepted letter from his wife, Jane, to a family member in Gonzales County: "On April 28, 1877, a day when Wes had been away for three weeks, Jane . . . wrote her uncle, Joshua Bowen, in Gonzales County."[9]

Jane indeed did write Joshua Bowen on that date. He wrote back to "Mrs Jane Swain"—the family alias at the time—on May 6, telling her of tragic news: "I recieved your welcom[e] letter of apri[l] 28th[.] I was more than glad to hear from you and to hear that you ware all well. . . . James Tayler [sic] was killed by the Suttin [sic] Party [and] Billy Tayler was re captured a few days scince and is in Austin jail[.]"[10] Neither Jane nor John Wesley Hardin could have imagined that these events were the beginnings of the law's serious efforts to rid the state of desperadoes such as the Taylors and, ultimately, John Wesley himself.

Among the best detectives in the state was John Riley "Jack" Duncan of Dallas. If he could ingratiate himself in Gonzales County, Hardin's home and the home of numerous relatives of his wife's, then perhaps the outlaw's location could be determined. Duncan was sworn in as a private in Lieutenant Hall's State Troops on July 15, 1877, his only assignment to work up the case with Armstrong.[11] Duncan, now "Mr. Williams," traveled to Gonzales disguised as a day laborer. He befriended Neill Bowen, Hardin's father-in-law, and claimed to be interested in renting a storehouse on the Bowen property. Bowen put Duncan off, saying he did not own the storehouse but would contact the owner. Bowen thought he had mailed the letter, but Duncan intercepted it and thus learned that Hardin was probably in Alabama.

This is only one version of how Hardin's location was learned. Hardin later claimed that a letter from Joshua R. "Brown" Bowen was intercepted which ended with the line that his sister sent her love. Bowen's sister was Jane Hardin, and Jane and John Wesley Hardin would be together, it was believed. Hardin wrote to Jane after his capture but still identified himself as John H. Swain: "Brown's Bad conduct caused me to get caught . . . in Pensicolia [Pensacola] and all so his Last letter to Texas Stating that his sister Joines Him in Sending Love to all[.] the Detective was Boarding at N. B [Neill Bowen's] when the Letter came an[d] watched them put the Letter away and then Stole the Letter out[.] N. B. thought the man Mr Williams to Be a merchant wanting to rent the Store Hous[e] But His name is John Dunkin [*sic*] a State Detective of Texas[.] Jane B. [Brown] is the cause of my arrest."[12]

On August 15, having received the notice from Duncan to "Come get your horse," the prearranged code that indicated that Duncan had the information he needed, Armstrong arrived with several others in Gonzales and "arrested" the transient Mr. Williams. The lawmen arrived in Austin on the seventeenth. Armstrong checked into the Avenue Hotel; either he or the clerk proudly wrote "Texas Ranger" after his name.[13] It was there that Duncan and Armstrong, with his cane, started to work on the Hardin case. A *Daily Democratic Statesman* reporter may have visited with him, for his arrival in the city was news. It was reported that Armstrong had "nearly recovered from a wound in the groin, caused by the accidental discharge of a pistol."[14] He asked Adjutant General Steele to cause warrants to be issued, one in Hardin's true name and another in the name of J. H. Swain, to be sent to him at Montgomery, Alabama. One was to go by regular mail and the other by express. Armstrong and Duncan left Austin on the train the next day. They were so anxious to get to the chase they left without any warrants or requisitions.

Two days later, on the twentieth, the Texas lawmen arrived in Montgomery. Armstrong now sent the first of seven telegrams to Texas. The first, dated August 21 from Montgomery, was sent to Steele: "Arrived last night[.] will wait here for papers[.] Duncan has gone ahead." Armstrong waited for the papers to arrive while Duncan went ahead to determine Hardin's exact whereabouts. He went to Pollard, Alabama, just north of the Alabama-Florida state line. Again pretending to be a transient, he contacted Escambia County, Alabama, deputy sheriff Neill McMillan and inquired about a fictional relative. Ironically, and certainly unknown to Duncan, Deputy McMillan's father, Malcolm McMillan, was the county sheriff and was married to a cousin of Hardin's wife. During their conversation Duncan casually mentioned Swain and was luckily told by Deputy McMillan that he lived in Whiting, a small community just north of Pollard. Duncan walked to Whiting, where he learned that Hardin was likely in Pensacola, Florida, on a gambling trip. He telegraphed Armstrong in Montgomery that their man was in Pensacola and he should meet him in Whiting.

Luck was with Duncan, as the railroad superintendent—William Dudley Chipley—took Duncan into his confidence, stating that Swain and his frequent companion Brown Bowen had made threats against him and were troublesome generally. He proved to be more than willing to help get them out of his territory. Armstrong joined Duncan on the morning of August 23 in Whiting. Chipley arranged for a special engine and car to deliver the two lawmen to Pensacola, Chipley accompanying them. They brought Escambia County sheriff William H. Hutchinson into their confidence. He was informed that they were to arrest John Swain. The lawmen learned that Swain intended to leave Pensacola on the afternoon train. Additional men were deputized, and Hutchinson's deputy, A. J. Perdue, was brought in as well. These temporary deputies were to be placed around the train station platform.

Hardin's capture has been described by a number of historians, all disagreeing in a detail or a major factor. What most likely happened is that Hardin, with friends Sheppard Hardy, Neal Campbell, and Jim Mann, and possibly one or two others, boarded the train at Pensacola to go to Whiting. They retired to the smoking car, as Hardin was now smoking a pipe. Curiously, they had shotguns, which they placed in the baggage racks above their heads. Armstrong, still using a cane but with a Colt .45 revolver as well, placed himself in the baggage car while Duncan remained outside on the platform. When the capture was complete Armstrong would signal the conductor to leave the station.

William Dudley Chipley, circa 1865. *Courtesy Lillian D. Champion and William S. Cummins*

As county sheriff, W. H. Hutchinson customarily walked through the cars to see if any undesirables were present, so Hardin took no special notice when he entered. Deputy A. J. Perdue also entered the car and even stopped to talk with Hardin about his staying over so Perdue could win back some money he had lost gambling. Hardin declined, indicating that he had other business to attend to before pleasure. With Hardin thus off his guard, Hutchinson and Perdue exited the car but returned immediately. Hutchinson stated to Hardin, "I believe I want you," to which Hardin replied, "Damn you, take that!" and kicked out, striking Hutchinson in the groin. Hutchinson and Perdue both grabbed Hardin, who was desperately trying to get his hands on his concealed pistol.

By now Armstrong had entered the other end of the car with a long-barreled Colt revolver in his hand. Hardin recognized the big weapon as typical of Texas and yelled out, "Texas, by God!" as he continued to struggle against Hutchinson and Perdue. One of Hardin's companions, young Jim Mann, either believing he was to be mobbed as well or simply wanting to get away, struggled to get out of the car. He jerked his pistol and fired "three fortunately harmless shots" but was shot for his troubles.[15] Perhaps it was Armstrong whose bullet killed Jim Mann, or possibly it was another's, as there were reportedly twenty shots fired in all, by Armstrong as well as the deputies outside.

Hardin related how it happened to his wife the first chance he had: "Jane they Had me foul yes very foul[.] I was Sitting in the Smoking car Neal C & Poor (Jimmie M) By my Side with my arms on the Sid[e] when they came in. 4 men grab[b]ed me one by each arm and one by each Leg. . . . But poor Jimme he Broke to run out of the cars and was Shot dead by some of the crowd on the out Side."[16]

Armstrong, clearly the man in charge of the entire operation, demanded Hardin's surrender. He replied, "Shoot and be damned, I'd rather die than be arrested."[17] As Hardin continued to struggle Armstrong hit him over the head so hard he thought he had killed him.

Hardin was delivered to Montgomery, where he was brought before Judge John A. Minnis on August 24. He employed attorney J. W. Watts to seek a release for him on a writ of habeas corpus. Opposing Watts was A. A. Wiley. Armstrong convinced the judge that his prisoner, known as John H. Swain, was indeed the notorious Hardin and that papers were on the way.

Armstrong kept Adjutant General Steele informed of events. On the twenty-third, at Whiting, he sent the second of his telegrams: "Arrested John Wesley Hardin Pensacola Fla this p m[.] He had four men with him[.] had some lively shooting[.] one of their number Killed all the rest captured[.] Hardin fought desperately [but we] closed in and took him by main strength[.] Hurried aboard the train then leaving for this place[.] we are waiting for a train to get away on[.] This is Hardens [sic] home and his friends are trying to rally men to release him[.] Have some good citizens with & will make it interesting[.]" The third telegram was sent on August 24: "Arrived this a m[.] prisoner in jail[.] no papers whatever received by the Governor[.] What is the matter[?]"

Armstrong showed the judge the papers authorizing him to make an arrest and convinced him of Hardin's character. Hardin remained in jail. Armstrong then sent Governor Hubbard the following: "To His Excellency R. B. Hubbard of Texas[:] Please telegraph the Gov'r of Alabama that you have forwarded requisition for John Wesley Hardin alias John Swain[.] They are trying to release him on writ of habeas corpus[.]" The last telegram, dated August 24, reads: "If requisition don't come tonight Gov Houston will issue a warrant on Gov Hubbards telegram so I can leave here at six tomorrow morning[.] Have arranged to have Bowen captured[.]"

On the twenty-fifth Armstrong sent a final telegram to Steele: "It is all day now[.] On our way[.] papers OK[.]" The arrival of the requisition from Hubbard was indeed timely as it enabled Armstrong and Duncan to start on their way with the "Grand Mogul of the Texas desperadoes."[18]

Armstrong had attempted to arrest Hardin once before. Relating the incident years later, he noted that he had heard that Hardin was in a small town in southwestern Texas, possibly Goliad, and learned that Hardin was on a drinking spree and was shooting up the town. He had last been seen entering a certain saloon from which other customers had quickly exited. Armstrong talked to a local law enforcement officer, and it was agreed that the two would attempt the arrest. The plan was for Armstrong to walk up to the bar, invite Hardin to take a drink, and while Hardin was thus engaged, each man would grab an arm,

The capture of John Wesley Hardin. From Hardin, *The Life of John Wesley Hardin*

the local lawman the right and Armstrong the left. Armstrong quickly had his man covered, and the man was jailed. It was to the lawmen's embarrassment that the "bad man" was simply a drunkard who was masquerading as Hardin. It was exciting, although disappointing, to see some action in going after the notorious Hardin. This led to Armstrong's asking the adjutant general to give him a special assignment to work up the case against the genuine Hardin.[19]

Armstrong recalled some of the details of the Hardin capture differently from what was reported in the press or related to others. Regarding the letter which was the key to locating Hardin, Armstrong's son Thomas Reeves, says, "When this letter was posted, Armstrong went to the Postmaster and asked to see it. The Postmaster refused to give it to him and he had to take it virtually by force. In this way they learned the whereabouts of Hardin, and the name which he had assumed, which was John Adams [sic], as I recall it." Thomas Armstrong preserved the story by relating it to his brother Charles Mitchell.

As Ranger Armstrong recalled, Hardin and five or six of his gang were in the car. Duncan pointed out Hardin sitting in a seat next to a window on the station side. He was facing forward with his right elbow resting on the sill of the open window. "The plan Armstrong outlined was that, since they intended to take him without any papers and they might have to do some shooting, he would enter the front of the coach and draw the fire; Duncan was to go up the side of the train and seize and hold Hardin's right arm which was resting on the open window-sill. The sheriff and his deputy were to enter by the rear of the coach and overpower him."[20]

In spite of the numerous and somewhat conflicting accounts of the capture of this noted desperado, one point that has been ignored by historians is the following: all the lawmen involved were certainly aware that the possibility, in fact, the probability, existed that there would be shooting. They were after the most dangerous gunfighter the State of Texas had produced, yet they were setting themselves up to be victims of their own "friendly fire," as Hardin would be between the two groups. With lawmen at each end of the train car converging on Hardin, how could they hope to avoid getting shot? As it was, Armstrong was the only lawman who fired his pistol—at least inside the car—but had he missed Mann, his bullet could easily have taken out Hutchinson or Perdue.

As Armstrong reached the platform of the coach, he shifted his cane from his right hand to his left and drew his Colt .45 Frontier model with its seven-inch barrel. When Hardin saw this he exclaimed, "Texas, by God!" and reached for his own pistol. Armstrong recalled that Duncan did not fulfill his part of the plan, which was to grab Hardin's arm through the window. Perhaps from the making of the plan to the point of being able to carry it out Hardin moved his arm. Certainly he did when he saw Armstrong's pistol and attempted to reach his own. Armstrong hastened down the aisle, probably not too quickly with his game leg, ordering the gang to surrender. All held up their hands except Hardin and Mann. Hardin's pistol was caught in his suspenders, but Mann had his out and fired at Armstrong. The bullet went through his hat and grazed his scalp.[21] It was sheer chance that Armstrong was not killed by this lucky shot. He responded the only way he could, by firing back, and Mann was killed with a bullet through the heart. Strangely, Mann did manage to exit the train but dropped dead on the station platform. It was also sheer chance that Hardin's pistol was caught in his suspenders. Did he have the suspenders looped through the trigger guard, or was the hammer caught in the suspenders? If indeed he had suspenders going through the trigger guard he had placed himself at a definite disadvantage, as even an amateur man hunter would then have the edge. Within seconds Mann had fired and missed Armstrong. Armstrong fired and mortally wounded Mann while Hardin struggled with his pistol and suspenders. The seconds Hardin wasted in attempting to get his pistol into action allowed Armstrong to do what he had set out to do: capture the man killer alive. Armstrong knocked him unconscious for several minutes.

Armstrong took the pistols from the rest of Hardin's entourage and ordered the train to proceed. At stops along the road, because he had no papers for them and because he had no real interest in such "small fry," he allowed the others to get off the train, with their pistols, but now emptied of cartridges. At one point a group of Hardin's friends was ready to take him by force. Armstrong informed them that if they tried to take his prisoner he would first shoot Hardin and then fight them as long as he had cartridges to fire. The mob quickly dispersed.

In court Armstrong explained the situation and stated that the papers for Hardin's arrest were on their way by mail and express and he had a telegram to that effect. "The only official paper that he had was his commission as an officer of the Rangers, which he showed them. He described the character and reputation of the famous outlaw and requested that they keep him in custody to allow reasonable time for the necessary papers to arrive. This the court decided to do." [22]

Armstrong wore a goatee at the time. "As the court adjourned, Hardin's wife, who had been in court, hastened up to him and grabbed him by the goatee and denounced him for having captured her husband. She showered him with abuse and said that she was going to raise her son to kill him. He said that was one of the most painful experiences he ever had." [23] Curiously, Jane Hardin's temper received some attention in the *Montgomery Advertiser and Mail* after the arrest, which corroborates Armstrong's recollection of her pulling his goatee. She was interviewed holding her six-week-old baby, Callie Jane, in her arms. Mrs. Hardin "has the bearing and converses like a person of much more than ordinary nerve and courage. She boasts of being able to shoot and manage a horse as well as most men, and says things will be made extremely lively for Armstrong and Duncan, the detective, and also for some others who had a hand in the capture of Hardin." [24]

Hardin cursed Armstrong all the way back to Texas. At all the stops for meals along the way, people stopped to see Hardin and the man who had captured him. Local authorities had been alerted to keep the crowds back at the various stations. When they got to Austin, Armstrong, fearing there might be an attempt to rescue Hardin, had a coach wait for them on the opposite side from the station, near the rear of the train. He hustled his prisoner into it and galloped the horses up Congress Avenue to the jail near the capitol, where he delivered his prisoner. No doubt, Armstrong and Duncan breathed a huge sigh of relief when their prisoner was finally turned over to jail authorities. [25]

On the morning of the twenty-fifth the press released an important news item, published not only in the *Atlanta Daily Constitution,* the *Chicago Tribune,* and the *New York Times* but also in numerous Texas newspapers as well as in those in other states. It was the first published account of the arrest: "Whiting, Ala., August 24—To-day [*sic,* yesterday] as the train was leaving Pensacola the sheriff, with his posse, boarded the cars to assist two Texas officials to arrest the notorious John Wesley Hardin, who is said to have committed twenty-seven murders, and for whose body $4000 reward has been offered by act of the Legislature of Texas. His last murder in Texas was the killing of the [deputy sheriff] of Comanche County, [Brown]. He has lived in Florida as John Swain, and being related to county officers, has escaped arrest. Twenty shots were fired in making the arrest. Hardin's companion named Mann, who had a pistol in his hand, was killed." We do not know whom the reporters talked with in gathering their details. Most likely, it was either Armstrong or Duncan, as it is doubtful that any of the Floridians involved would have been aware of the number of Hardin's alleged victims.

Even in irons, Hardin did not go meekly. While in the Memphis jail it was discovered that he had a knife up his sleeve, which Duncan took from him. This demoralized Hardin considerably. Duncan reminded him that, if a mob should attack with the intention of killing him, he would be given pistols to help defend himself, and the three of them would fight off the mob. This appears only in Duncan's account, not in Armstrong's.[26]

In his autobiography, Hardin recalls that at Decatur, Alabama, when his handcuffs were removed to facilitate eating he noticed Armstrong's pistol partly exposed. Duncan's quick action prevented him from getting his hands on that pistol. Hardin writes that he knew his only hope was to escape: "My guards were kind to me, but were most vigilant." He promised to be quiet and thus "had caused them to relax somewhat and they knew that their life depended on how they used me." At Decatur, as Hardin recalls, it was necessary to stop and change cars for Memphis. "They took me to an hotel, got a room and sent for our meals. Jack and Armstrong were now getting intimate with me, and when dinner came I suggested the necessity of removing my cuffs and they agreed to do so." As Armstrong unlocked the handcuffs and started to turn around, his six-shooter was exposed, but "Jack jerked him around and pulled his pistol at the same time."

"Look out," Duncan said. "John will kill us and escape." Hardin claims to have laughed at him, but "Jack had taken the play away just before it got ripe. I intended to jerk Armstrong's pistol, kill Jack Duncan or make him throw up his hands." Hardin explains that he could then have got the key, which would enable him to escape easily. "That time never came again."[27] No doubt, Armstrong was very careful to keep their prisoner alive, as a dead Hardin would not have brought them the $4,000 reward offered only for his delivery inside the doors of the Travis County jail. He had not forgotten his ill luck with the killing of Mayfield in Wilson County.

The Rangers did not cross the Texas line until August 27. Crowds gathered frequently along the way, anxious to see the noted desperado. On the morning of the twenty-eighth the trio arrived in Austin. Upon their arrival in Austin Armstrong and Duncan learned that there was "a tremendous crowd at the depot," so they managed to stop the train and acquire a hack to take them directly to the Travis County jail. The crowd at the depot learned of this and "broke" for the jail, not wanting to be denied seeing the noted celebrity. "The hack just did manage to get there first and they carried me bodily into the jail; so when the crowd arrived, they failed to see the great curiosity." Hardin was placed in the new Travis County jail.[28] Only three years earlier that jail had been considered one of the most horrible institutions in the state because it was void of adequate ventilation, and prisoners sweltered in the heat and the stench of their own waste.

At first Hardin, while speaking with a reporter from the prestigious *Galveston Daily News,* claimed to have been kidnapped and that he did not know what charges were being brought against him. He also indicated that he was not afraid of going to Comanche except for the possibility of mob action against him.

In the Travis County jail were some seventy or eighty prisoners. Some Hardin knew from past actions. Bill Taylor was there, now incarcerated for his part in the killing of Sutton and Slaughter back in 1874 during the climactic years of the Sutton-Taylor feud. Hardin's first cousin Mannen Clements was there, also charged with murder. Jefferson Ake and John Ringo were there. Ake would gain some recognition by writing his life story, while Ringo would gain fame in Tombstone in conflict with Wyatt Earp. George Gladden was there as well, having been found guilty of murder during the "Hoo Doo" War in and around Mason County. Sam Pipes and Alf Herndon had been incarcerated for their participation with the Sam Bass gang of train robbers.

Hardin was scarcely behind bars before efforts were made to round up his brother-in-law, Brown Bowen. Five days after Hardin's capture, William D. Chipley penned a letter to Governor Hubbard from his office in Pensacola. He reminded the governor of his involvement in Hardin's arrest and now urged him to offer a suitable reward for Bowen. "Lieut Armstrong will explain to you, that we may have to buy him. I have a party regularly employed & am willing to do everything in my power but am not able to stand it all," explained Chipley. He continued, stating that he had received papers that very day for Bowen's arrest in Alabama but feared he would get away and flee to south Florida: "Bowen tried to assassinate me while on my depot platform[,] a perfect stranger to him. . . . I nearly Killed him at the time with the punishment I gave him. I want you to help me. . . . Lieut Armstrong assured me that a reward would be offered & I have held out such an inducement to the men who are on his track." [29]

After reading the letter Governor Hubbard made a note in the margin to telegraph a reward of $500 for the capture of Bowen. On September 17 Bowen was indeed taken near Pensacola and sent to Texas in the custody of Florida officials, probably Hutchinson and Perdue. Presumably, Chipley received the $500 reward for Bowen's capture.[30] Bowen was charged with the murder of Thomas Haldeman in December of 1872, a crime which he accused Hardin of committing. A Gonzales County jury found Bowen guilty of murder on October 18, 1877, and on May 17, 1878, he was legally hanged in Gonzales.

Hardin ultimately stood trial for the killing of Charles Webb, was found guilty of murder in the second degree, and received a sentence of twenty-five years to be served in Huntsville State Penitentiary. He was pardoned in 1894 and, after drifting westward to El Paso, was killed there on August 19, 1895.[31] Jack Duncan continued as a detective for many years, although no other case brought him as much fame as did his involvement in the Hardin capture. He died as a result of an automobile accident on November 16, 1911.[32]

Sheriff William Henry Hutchinson believed he had been unfairly treated by the Texans and as late as 1895 was making his complaint public. According to his version he and his deputy had Hardin totally secured by rope and Mann was shot and killed by someone from the outside, *before* Armstrong or Duncan even entered the train car. Hutchinson claimed that he was "fully cognizant of who Hardin was, as well as of his history in the past and what kind of man he had to deal with, as I went directly opposite to the ideas of Lieut. Armstrong and Duncan in making the successful move culminating in the arrest, while they remained on the platform of the depot." Hutchinson recalled that he and Perdue took Hardin to the state line and there delivered the prisoner to Armstrong and Duncan. There, Hutchinson shook hands with Hardin, saying he "regretted the necessity" of arresting him. Hardin responded, "I have killed twenty-seven men, and Hutch, you came near being the twenty-eighth." If indeed Hutchinson was "fully cognizant" of who Hardin was and was aware of "his history in the past" and "what kind of man he had to deal with," why had he not arrested Hardin prior to the arrival of the Texans? Hutchinson's claim does not remotely ring as true.

Hutchinson further complained of the amount of the reward he received, saying he was the "recipient of a paltry $500, which was given to those assisting me in the capture outside of Lieut. Armstrong and Duncan."[33] It is difficult to accept anything of Hutchinson's account as being accurate. He certainly was involved but by placing Armstrong and Duncan totally out of the actual capture —when they were the ones who made it succeed—he throws his entire account into doubt.

Hutchinson was superintendent of Pensacola streets when death claimed him on January 14, 1911, at his home.[34] A. J. Perdue, who apparently lost money gambling with Hardin, died before 1895, although neither the exact date nor the circumstances are known. William Dudley Chipley continued his long career in public service, serving as mayor of Pensacola, a member of the Board of City Commissioners, vice-president of the State Democratic Executive Committee and chairman from 1888 to 1890. He died on December 1, 1897. A tall monument was erected to his memory in the public square of Pensacola. The county seat of Washington County, Florida, is named Chipley in his honor as well.[35]

It is perhaps appropriate here to attempt to dispel several inaccurate myths about the Hardin capture. An early myth originated in the work of N. A. Jennings, who claimed to have interviewed other Rangers in writing his memoir, *A Texas Ranger*. Jennings had joined McNelly's company and served from May 26, 1876, until February 1, 1877; thus he was no longer with the company when Armstrong made the arrest. Jennings wrote that Jack Duncan, after determining Hardin's probable whereabouts through the intercepted letter, "went straight to San Antonio, where he met John Armstrong, Charley McKinney, and two other Rangers. The party at once started for Florida." Curiously, Jennings makes no further mention of the other Rangers by name or of their involvement in the capture, but does write, in reference to the reward, that "Armstrong and his *four* companions received $800 each from the State for capturing Hardin" (emphasis added). He further states that the lawmen later regretted that Hardin had not been turned over to authorities in Louisiana, "for there was a reward of $12,000 for him in Louisiana, for murders committed there."[36]

Almost the entire account by Jennings must be rejected for a number of reasons. First, as far as is known, Hardin had no real trouble in Louisiana. No mention is made of any difficulty in his autobiography, and no contemporary record has been located. Further, Jennings mentions Armstrong's four companions, but his identification of one as Charles B. McKinney is incorrect, as McKinney had nothing to do with the capture. Could Jennings have intended to identify the four companions as Duncan, Hutchinson, Perdue and Chipley? Finally, it is unknown how Jennings determined, or possibly heard, that the captors received $800 each for their part in the capture. Or did he just make an educated guess that, if the five were involved, then the $4,000 reward must have been divided equally, each receiving an equal part? This does not seem plausible, as, after all, Hutchinson himself claimed he had received but a "paltry $500."

Although there are too many writers who have added their own "insights" to the capture of Hardin to itemize here, one of the strangest accounts appeared immediately after Armstrong's death in 1913, in the *San Antonio Daily Express*. I suspect that a reporter tried to get information from a family member at Armstrong's funeral and was rebuffed, hence the need for "creative journalism." The reporter wrote that Armstrong wanted to go after Hardin alone, "for fear that others might bungle" the attempt. He found, however, that it was necessary to have someone with him "to attract Hardin's attention when necessary. He found in Florida a man who knew Hardin unmistakably, but whom Hardin did not suspect." The pair found Hardin standing by a well and engaged the desperado in conversation. Armstrong's companion "suggested that Wes draw up a bucket of water, as he was thirsty. Hardin set his rifle down at his right, caught the well rope in both hands and commenced to lower the bucket. While he was leaning over the well Armstrong poked two pistols under Hardin's nose, as he kicked the rifle to one side." This account is so filled with errors of omission and commission that it is needless to comment further.[37]

In 1952 Hollywood released a motion picture purporting to tell Hardin's life story, in which Armstrong makes a brief appearance. *The Lawless Breed* is based very loosely on Hardin's autobiography. In the film Armstrong first appears as a member of McNelly's Ranger force hunting Hardin shortly after the Webb killing in 1874. Armstrong has tracked his prey to Kansas City and sent a note, signed simply "Armstrong," to McNelly that he has located the fugitive. Even with the additional reinforcements Hardin manages to shoot his way out of the carefully laid trap, although Armstrong ultimately does track him to Pensacola, where the arrest is made *inside* the railroad depot.[38]

With the most desperate man in Texas behind bars, Armstrong soon had other thoughts occupying his mind, one of which was how best to invest his share of the reward for Hardin's capture. Other thoughts concerned Miss Mollie Durst of Austin.

Ranger Armstrong was present at the death of outlaw Sam Bass. Bass is shown here with three friends in a picture believed to have been taken in Dallas in 1876. Standing, left to right: Bass and John E. Gardner. Seated, from left: brothers Joe and Joel Collins. *Courtesy Western History Collections, University of Oklahoma Libraries*

Action in Round Rock

Lieut. Armstrong took away the charmingest girl from the capital.

—*DAILY DEMOCRATIC STATESMAN*, FEBRUARY 26, 1878

T HERE is considerable question over the matter of the reward for John Wesley Hardin and how it was distributed. The state offered $4,000 for the desperado's delivery to the Travis County jail. Who actually received the reward or even whether the entire amount was paid is not known. Jack Duncan, in discussing his career with a Dallas newspaper reporter years later, listed some twenty-one fugitives he had captured, naming Hardin first. He stated that he had received $4,000 for Hardin, although he certainly could not have been given the entire amount.[1] Curiously, Duncan, in an interview given only days after the capture, declared that Hardin's "captors" were paid $200 per month by the state to work up the case and $5,000 reward besides.[2] If this is true, then Gov. Richard B. Hubbard made special arrangements with Duncan and Armstrong which never received publicity in the press, nor has any official document been discovered verifying this. Armstrong—for the June 1–August 31 pay period—received $375, the standard amount for a lieutenant in the Special State Troops.[3]

Before the Rangers were even back in Austin there was talk on the streets that Armstrong would be a good candidate for the position of city marshal. The *Daily Democratic Statesman* learned of the talk and commented that because of his "gallantry and daring and good judgment in arresting desperate men" it was only natural that his "many friends" in Austin would ask him to allow "the use of his name for that office."[4] Armstrong no doubt could easily have handled the position of city marshal with great skill; he rejected the notion probably because of the politics he knew would be involved. Like the requests that he run for sheriff of several other counties, Armstrong apparently did not give this talk serious consideration.

While Armstrong and Duncan were working up the case against Hardin there was at the same time a plot developing by various fugitives and convicts to eliminate certain lawmen, including Armstrong. Whether Armstrong was ever informed of it is not known. We are aware of it only through a statement given by convict W. A. Bridges to Huntsville State Penitentiary superintendent Thomas J. Goree, who recorded the statement and sent it to Adj. Gen. William Steele.

Bridges told Goree that while he was in the Travis County jail a letter was delivered to the cell occupied by Jack Jones, John West, Bill Williams, and Bill Allison. It was addressed simply "Dear Friend" with no signature, but Jones knew who it was from. The letter in essence explained how a "band" was being formed and would meet soon near the Frio Canyon. The outlaw group would get Mexicans disguised as Indians to "depredate on the extreme frontier, to engage the attention of the State Police & Rangers." Although not stated, it was understood that while police and Rangers were occupied with the "Indian menace" the "band" would release prisoners from jails in western Texas. Once this was accomplished they would then liberate prisoners in the Travis County jail and, if successful there, attack the state penitentiary. The ultimate goal was to have sufficient force "to plunder towns, banks, &." [5]

But in addition to plundering, certain individuals were to be killed. The first name on the list was that of Maj. John B. Jones, head of the Frontier Battalion. Armstrong was to be assassinated, as well as Hall and Lt. John C. Sparks of Company C, Frontier Battalion. There were spies in Austin who would obtain information about the disposition of Rangers and police at the proper time.

Needless to say, this grandiose plan never caused serious concern on the part of law enforcement, although there may have been some investigation not reported or documented. It is doubtful that Armstrong or the others really considered it a grave threat, as, after all, their job was to face danger and possible death on a daily basis. Rumors were already circulating about an attack on the Travis County jail. According to one report, the people of Austin were worried about "a general revolt and an outside attack upon the jail, which contains over seventy prisoners, the majority of whom are among the worst desperadoes known in the criminal annals of Texas." It was believed that Hardin's gang would lead the attack to release him.[6] No attack was made, although certainly at least some of the prisoners were continually thinking of escape plots.

Armstrong may have had a brief period to relax after bringing back Hardin, but among his first responsibilities was to deliver the notorious Bill Taylor to trial. Taylor, accused of the 1874 killing of Gabriel Webster Slaughter on the deck of the steamer *Clinton* at Indianola while cousin Jim Taylor allegedly killed William E. Sutton, had been captured. His trial was to be held in Indianola, Calhoun County, and Armstrong with a squad of six men delivered him there without incident.[7] Ironically, only days before, on October 18, Brown Bowen had stood trial and was found guilty of the murder of Thomas Haldeman. He was sentenced to be hanged in Gonzales, Gonzales County.

The Sutton-Taylor feud provided more work for the Rangers, as Lieutenant Hall had arrested the seven men charged with the double murder of Dr. Phillip H. Brassell and his son George. The DeWitt County jail was considered insecure, and they were therefore confined in the Travis County jail until their trial. The seven—David Augustine, William Meador, Jake Ryan, Jim Hester, William Cox, Joe Sitterle, and Charles Heissig—were delivered to Cuero by Armstrong, arriving there on Sunday, December 16. To the people of Cuero, who had endured years of strife between the Taylor and the Sutton parties, the belief that this was "one of the most important trials that has ever taken place in this district, and has attracted a good deal of attention throughout the State" was certainly an understatement.[8]

One of the Rangers making up Armstrong's squad left a fascinating account of his experience, although his name is unknown, as he signed his correspondence simply "Total Wreck," a nom de plume possibly inspired by being continually exhausted from hard rides and lack of sleep. Once in Cuero he had the time to record his experiences:

After leaving Corpus we were stationed at Banquette [present-day Banquete], the present terminus of the Corpus Christi and San Diego Railroad. Finding after staying about two weeks that we could not accomplish anything at this place, Lieut. Armstrong sent a squad of us to San Diego where we remained for three days. Our orders were to assist the sheriff of the county in the discharge of his duties. After our arrival we were kindly furnished quarters for the night by the sheriff to await the arrival of Lieut. Armstrong, who arrived the next day from Corpus. Finding peace reigning supreme, there not being a criminal case on the docket, consequently we had to again take to the saddle for Corpus, much to our delight. We only stopped in Corpus one day when we were told to get ready for a long march, whither bound we knew not, but some of the shrewd ones said to Austin, which turned out to be correct. We were ordered there for the purpose of escorting some prisoners to Cuero, who were to be tried for the murder of the Brassells. We arrived in Austin on the 12th inst., tired and the hungriest set of men that the Hill City ever entertained.[9]

While in Austin Total Wreck had his curiosity satisfied by making the acquaintance of Hardin in the county jail. Hardin proved to be "a great talker" who "speaks in high terms of his captor," whom he considered to be "a Texan of the old school." Hardin claimed he "was never treated better by anyone in his life," certainly high praise for Armstrong. Hardin compared him to the various other lawmen in whose hands he had been, although in the previous cases where he was a prisoner he had always managed to escape. Total Wreck points out that Hardin and his brother-in-law, Brown Bowen, occupied the same cell and although Bowen had accused Hardin of murdering the man for whose death he was to be hanged, they both seemed "to be in the best of spirits."

On the road to Cuero the Rangers and their seven prisoners stopped to spend the night in Lockhart, the county seat of Caldwell County, their sleeping quarters set up in a blacksmith shop. During the night the group was disturbed by the sound of several pistol shots. "Hastily giving the alarm, we were soon under arms, and headed by our gallant lieutenant, bound for the scene of the nocturnal disturbance," writes Total Wreck. They were "so fortunate" as to round up three of these "midnight angels and two pretty nickel-plated revolvers"; the prisoners were placed in the Caldwell County jail.

Armstrong and his prisoners left Lockhart at 4:00 PM, their next stop the town of Gonzales, five hours and thirty-five miles later. They spent the night there, apparently with no disturbance, and arrived in Cuero the next day. Cuero was the new county seat of DeWitt County, having replaced Clinton in 1876 with the arrival of the railroad.

On entering Cuero the Rangers and their prisoners were met by "a whole delegation of ladies, the wives, sweethearts and relatives of the prisoners, and what a joyful meeting it was" writes Total Wreck. "I was almost sorry I was not a prisoner instead of one of the guards. We have been attending court here ever since our arrival, and growing almost as impatient as the prisoners themselves for a trial, as we are sadly used up for want of sleep, being on guard every night, not having men enough for a change of guards." Total Wreck was doubtful the prisoners could get a fair trial in DeWitt County and surmised that there would be a change of venue. He concludes, "This is their third trial."

Hardin's capture had catapulted Armstrong into fame, making him well known across the State of Texas. Not only did it bring him unwanted recognition but it also resulted in his being promoted to first lieutenant, his new rank taking effect on December 1, 1877.[10] A visit to Austin almost guaranteed at least a mention in the newspapers. The new year of 1878 began with this notice in the *Daily Democratic Statesman:* "Lieutenant Armstrong, of the State Troops, is in the city."[11] Of course, there was no need to explain who Armstrong was. A few weeks later Armstrong was in Victoria County, southeast of DeWitt County. It was noted in the *Victoria Advocate* that he, a brother of Capt. J. L. Hall, and a private of the company "passed through town on Monday on their way to Cuero."[12]

The lieutenant was no longer hunting fugitives, although he was still a Ranger. His visit to Austin this time was much more personal: her name was Miss Mollie Durst. With a solid reputation for bravery and courageousness, admired and respected in Austin by the best of society for his work with the Travis Rifles, tall, handsome, and confident, he now felt the time right for courting the young lady for whom he had held strong feelings for several years. And Armstrong was successful in his suit, as on Wednesday, February 20, 1878, at 8:30 PM, John B. Armstrong and Miss Mollie H. Durst were married at St. David's Church in Austin, the Rev. Thomas B. Lee officiating. Attending the ceremony were some of Texas' finest: Captain Hall of the Special State Troops and Miss Pattie Townes; Thomas Netteville Devine and Miss Ella Mabry; Lieutenant McMurray and Miss Florence Jackson; Sterling F. Grimes and Miss Bettie Johnson; Robert J. Kleberg and Miss Julia Pease; Alex M. Jackson and Miss Ella Carter.[13] A "large congregation" of ladies and gentlemen were present to witness the ceremony, "impressively performed by the Rector, and to gaze upon the wedding party." The *Daily Democratic Statesman* described Miss Durst as having been for a long time "a favorite in Austin, not only with young gentlemen, but with our citizens generally, and it was natural that much interest should be felt in her marriage." As she leaned on the arm of the "gallant Lieutenant, all eyes were upon her, and many hearts wished her much happiness." Following the ceremony, the wed-

Mollie Durst about the time of her marriage to John Armstrong. *Courtesy Tobin Armstrong*

ding party and invited guests retired to the residence of the bride's mother, where a "fine array of costly presents greeted the eyes of the bride when she entered the parlor."[14]

The *Daily Democratic Statesman* played up Armstrong's celebrity status. In the issue published on Friday, February 22, it listed many of the gifts the couple received and who gave them. Armstrong's new mother-in-law presented them with a silver cake stand and a set of silver knives. Concluding its report the paper declared that there were "numerous other very handsome presents, perhaps more than was ever bestowed on any bride in this city." To demonstrate their generosity, the Armstrongs in turn gave to the paper a salver of "nice cake and wine."[15] The *Daily News* of Galveston also reported the "great event of the social season" in Austin and also listed some of the guests and what they gave. Captain Hall presented the couple with a "handsome pair of gold bracelets" and perhaps joked with the groom that they could be used as handcuffs.[16]

Four days later the Armstrongs were still newsworthy. In the *Daily Democratic Statesman's* popular "Texas-Facts and Fancies" column this notice appeared: "The Goliad *Guard* [of Goliad County] is sad because Lieut. Armstrong took away the charmingest girl from the capital." Later in the same column it was reported that there was a movement in DeWitt County to make Armstrong the "high sheriff." Smirked the reporter, "Everybody in these parts will vote for one-half of him."[17] Apparently, this was a serious effort on the part of the citizens of Cuero; after all, the man who could bring in John Wesley Hardin certainly would not allow any disturbance on the town's streets. The people of Cuero and Clinton and everywhere else in DeWitt County had had enough of fighting during the Sutton-Taylor troubles. The news reached as far north as Dallas, for as late as April the *Dallas Daily Herald* could report that one hundred citizens of DeWitt County had called on the Ranger, who "consented to become a candidate for sheriff at the next election."[18] But this might have been only wishful thinking on the part of the Dallas press, for Armstrong was not really interested in becoming sheriff of DeWitt or any other county. Although he remained in the Ranger service for another year, he was also contemplating land development and ranching.

Armstrong and his bride were again reported in Austin in April.[19] But he had to get back to rangering so left his bride with her mother. He went to Cuero to subpoena some witnesses in the Brassell murder case and safely delivered them to San Antonio, where the trial had been moved. He arrived there on April 12, delivered the witnesses, and then checked into the Menger Hotel.[20]

On May 25 Lieutenant Hall, Armstrong, "and a lot of the State Troops" were in Victoria.[21] Reports of Armstrong's alleged interest in a sheriff's position still floated about. The *Daily Democratic Statesman* reported in late June that Armstrong was running for the position of sheriff of Guadalupe County, "and we are glad to know with good assurance of success."[22] Whether Armstrong was ever really interested in running for sheriff of any county is unknown; he never became an official candidate for sheriff of any Texas county.

With his capture of John Wesley Hardin and his involvement in arresting John King Fisher, perhaps it is not surprising that Armstrong was present at the finale of yet another noted Texas desperado—Sam Bass. The Texas folk hero gained popularity and notoriety by robbing stagecoaches in the Dakota Territory and then making one big haul with the robbery of a train at Big Springs, Nebraska, in 1877. Following this successful robbery Bass and his gang continued with train robberies, only now in northern Texas. Never again did the gang achieve the success of Big Springs, however. On July 19, 1878, Bass and two others, Frank Jackson and Seaborn Barnes, went into Round Rock, Williamson County, a few miles north of Austin, to look over the town once more prior to a robbery planned for the next day. What Bass and his men did not know was that a traitor in the gang, Jim Murphy, had alerted Major Jones of the Frontier Battalion of the plan, and Jones was there with other Rangers intending to capture them in the act. Major Jones anticipated that he would need more men than he had and telegraphed Captain Hall, then in the capital attending the state Democratic convention, to come to Round Rock. Hall arrived in town early that afternoon and decided to telegraph Armstrong to also come to Round Rock with additional men.

However, Williamson County deputy sheriffs Ahijah W. "Caige" Grimes and Maurice Moore observed Bass and his companions openly wearing their pistols, which was against city ordinance, and challenged them, not realizing they were the Bass gang. Instead of surrendering their weapons Bass and Barnes drew and fired, and the gunfight began. Barnes was killed immediately, Bass was wounded, and Deputy Grimes was mortally wounded as well. Barnes was left dead in the street; his companion, Frank Jackson, managed to get Bass on his horse and out of town before leaving him to die. Jackson was never apprehended for his part in the various robberies. The gunfight in the streets occurred before Armstrong could arrive, so he missed out on the dramatic battle ending the career of outlaw Sam Bass. Bass was captured the next day by a squad of Company E led by Sgt. Charles L. Nevill.[23]

The mortally wounded Bass was brought into Round Rock and questioned extensively about his crimes and companions. What he admitted to was very little and not helpful to the authorities. It was perhaps the reporter from the *Galveston Daily News* who provided Major Jones with a notebook in which Bass's statements were recorded. Bass said of the street fight, "Grimes asked me if I had a pistol. Said I did, and then all three of us [Bass, Barnes, and Jackson] drew and shot him. If I killed Grimes it was the first man I ever killed." Armstrong was present, along with several others, during this questioning. "Have been in the robbing business a long time," said Bass. "Had done much business of that kind before the U.P. [train] robbery last fall." At this point the reporter noted that these latter statements were made in the presence of Major Jones, Hall, Armstrong, Rangers Henry McGhee and Richard C. Ware of Lt. N. O. Reynolds's squad, and Dr. C. P. Cochran.[24]

Considering all accounts, this is the most accurate explanation of why Captain Hall and Armstrong were in Round Rock that day. Strangely, Hall left a version of the end of the Bass gang which is quite at variance with the official record. He provided a lengthy interview with a reporter from the *San Antonio Daily Express* in September 1895. Perhaps because of the passage of so many years, his memory was truly faulting him, rather than his intentionally attempting to distort the factual record. Hall declared that the traitor Murphy had written to him, as well as to other individuals, telling of Bass's movements. Hall intended to meet with Murphy in Austin but Murphy was not there. "At this time the convention . . . was in session. Not meeting Murphy I remained in Austin in attendance on the convention, and was elected sergeant-at-arms of that body. In the meantime Gen. Steele sent for me and told me that he had had an interview with Murphy, who told him that Bass was in or about Round Rock, where he was planning a raid on a bank. Gen. Steele told me to go immediately to Round Rock . . . with a squad of men."[25]

Hall then states that he left "immediately" for Round Rock, taking with him his brother, Richard M. Hall, Lieutenant Armstrong, T. Netteville Devine, and James E. Lucy. "I left them, however, about three miles outside of Round Rock and went into town alone." He offers no explanation as to why he left them outside of town. Hall may have believed that what he was explaining to the reporter was accurate, but the official record shows conclusively that the gunfight was over prior to Hall's entry into town, contrary to his statement that he was in town prior to the gunfight. Further, Armstrong did not arrive at Round Rock with Hall but later the same day.

Armstrong continued to be important news as late as November 1878. His presence was noted in the *Daily Democratic Statesman* in a one-line item: "Lieut. Armstrong is in the city."[26] A year later his *absence* was noteworthy, for on October 1, 1879, it was reported that "Capt. John Armstrong" had returned the day before from a visit of several weeks to Washington, DC, and to Tennessee.[27] No record has been yet found explaining his presence in the nation's capital, but his trip to Tennessee was no doubt to visit friends and relatives, perhaps to introduce them to his bride.

John B. Armstrong, circa 1895. *Courtesy Tobin Armstrong*

CHAPTER EIGHT

Pioneer Ranchman

My time and work . . .
would be a fair off set against the expenses. . . .
Am confident we will win.

—J. B. ARMSTRONG

ARMSTRONG officially retired from the Texas Ranger service on December 30, 1878, receiving $133 for his final payment for his service with Hall's Special State Troops. He had memories of his time as a Travis Rifleman with bayonets flashing on Congress Avenue while protecting newly elected governor Richard Coke, the gun battles with bandits, recovering stolen cattle and horses, the excitement of the chase, and the capture or killing of wanted fugitives. The three-and-one-half decades remaining for Jno. B. Armstrong, as he now typically signed his name, were not filled with tediously preparing reports or filling out documents but with excitement of a different kind.

The new year of 1879 brought drastic changes in Armstrong's life. As a Ranger he could live on the ground, occasionally spending a night in a hotel, such as the Menger or the Central in San Antonio, but with a wife now he found residence with his mother-in-law on Hickory Street (present-day 8th Street) in Austin. In the early 1880s Mary J. Durst was head of a household consisting of her daughter and son-in-law and her one-year-old granddaughter, Maria Josephine, born April 5, 1879. The name given to the first Armstrong child carried on the first names of John's and Mollie's mothers, although she was always known as Josephine.[1]

Armstrong, besides being a husband with a young wife and child to love and support, entered a new career—that of real estate and land development. He worked as a clerk with Johns & Spence, proprietors of the Texas Land Agency, whose offices were at 1000 Congress Avenue.[2] His office, possibly not the first but where he was located during the 1880s, was then on the northeast corner of Congress Avenue and Bois d'Arc Street (present-day 7th Street), a mere four blocks from the capitol building itself. The letterhead announced "Office of Jno. B. Armstrong, Dealer in all kinds of Real Estate and Live Stock and Gen. Com. Agt."[3]

At first, until they could establish a home of their own, the Armstrong couple resided with Mollie's mother at her boardinghouse on the north side of Hickory Street between Brazos and San Jacinto Avenues. Two decades later, in 1900, the Armstrong home was a nearly palatial residence at 2610 Whitis Avenue. Their telephone number was 641.[4] This impressive residence was years in the future, but Armstrong now had a beautiful wife and daughter; they were acclaimed by society as an ideal couple; he was beginning a successful career without the dangers inherent in being a Texas Ranger.

Not all was serious work, however. In early June of 1882 Armstrong's mother came to Austin for a visit. Her arrival "a few days ago" to spend some time with her son and his family was noted in the *Austin Daily Statesman.*[5] Occasionally, Armstrong was called on to participate in civic functions. In August of that year he agreed to be a judge in a contest between two drill teams, the Broom Brigade (young ladies) and the Austin Greys.[6] Austin's Turner Hall was "literally packed, standing room even being hard to procure," according to the *Austin Daily Statesman.* Judges were Adj. Gen. Wilburn Hill King, "Captain" John B. Armstrong, and Col. Will Lambert.[7] Each company was composed of sixteen members. The Austin Greys drilled first, having substituted their rifles for the young ladies' brooms. "The boys drilled finely," smirked the newspaper, "in most respects, so far as the marching was concerned, but the manual of arms, with the brooms, puzzled them." They did the best they could and left the hall "amidst much appreciative applause." Then the young ladies entered and "went through the many intricacies of the drill with the most perfect precision." The judges were impressed, and it soon became "apparent to the general observer that the girls were the more perfect, and the decision of the judges in their favor was much applauded, not at all because of a desire to see the girls win merely, but from the universal and honest opinion that they did the finest drilling." Following this display, the Greys reentered and exhibited their abilities in the manual drill— with their guns. Then the girls came back, in their red jackets, and went through "other and more difficult evolutions." General King, with a few remarks, gave the unanimous vote of the judges to the female "soldiers."[8]

But there was something lacking: John Barkley Armstrong wanted a ranch to call his own. Surprisingly, owning a ranch was not out of the question as Armstrong soon learned that there was land in the Durst family. Mr. and Mrs. James H. Durst, Mollie's parents, had had a similar vision of owning land and a successful cattle ranch; now John B. Armstrong was in a position to turn that dream into a reality.

James H. Durst was born in 1819 in Nacogdoches, Texas, the son of Joseph and Delilah Dill Durst. In 1835 the census taker enumerated the Dursts in Nacogdoches County: Joseph was listed as a farmer/rancher, age forty-five; Delilah, age thirty-five, was keeping house; they had a fifteen-year-old son, James H. Both Joseph and Delilah died circa 1843–44 and were buried in eastern Cherokee County, Texas.[9] By the time James H. Durst was twenty he was captain of a Ranger company "who kept the Cherokees in check from December 1, 1838 to January 25, 1839." In 1840 he became commissioner in Nacogdoches County to detect fraudulent land claims. He married Tennessee native Elizabeth R. Culp on May 13, 1843. A son, Mortimer T., was the only child born to this union as Mrs. Durst apparently died in late November.[10]

Following the death of his wife James H. Durst left East Texas for Starr County in deep South Texas. Three days after Christmas, 1852, he purchased the lower fourteen leagues of La Barreta grant, now in Kenedy County. La Barreta (the barrier) originally comprised sixty leagues of land and was made out to José Francisco Ballí in 1801. In 1853 Durst was elected to the state senate representing Starr County, in which position he served until 1855. On January 11, 1854, he had married Mary Josephine Atwood at St. David's Episcopal Church in Austin, the same church where their daughter Mollie would later marry John B. Armstrong. Miss Atwood was also a Tennessee native.

Their first child, Mary Helena "Mollie," was born in Austin on March 22, 1855. When Durst's term ended in 1855 he and his wife returned to South Texas, where he became collector of customs at Port Isabel, Cameron County. A second child, James William, was born in Brownsville, March 28, 1857. Unfortunately, Durst died on April 24, 1858, on his ranch in Nueces County. The large property was claimed as his homestead. Upon his death the widow Durst, twenty-four years of age with two young children, relocated to Austin to live with her family.[11] She left thousands of acres of raw and untamed land in what is today Kenedy County, with no fences, no roads, no law, only roaming wild animals.

Exactly how Armstrong learned that there was land once owned by Durst but now "lost" is uncertain, but with his close relationship with his mother-in-law, known affectionately as "Granny" Durst, and his knowledge of land sales, it is not surprising that the previous actions of certain attorneys became suspect. James William Durst had sold certain land via three deeds executed in the settlement of James H. Durst's estate. John and Mollie Armstrong and her brother, James W., brought suit to cancel these three deeds, which had resulted in the illegal sale of that land, land which today makes up the Armstrong Ranch, as well as other land. James H. Durst died in 1858, and his will was probated the same year, but it was not until eleven years later, on April 22, 1869, that Mortimer T. Durst, Josephine Atwood Durst, and her children executed a power of attorney to two Nueces County lawyers—Lovinskiold and Campbell—to settle the estate. The power of attorney agreement provided the lawyers half of the money from the estate, as well as half of the land. On October 21, 1870, these lawyers filed an application to sell all of the estate's lands to pay the estate's debts and to give Josephine an allowance. In February 1871 some seventy-seven thousand acres of land were sold for a total of $3,737.31. Nearly thirty-one thousand acres of the Barreta grant were then sold to Richard Jordan for $697.84, or one and one-eighth cent per acre. On March 23, 1875, Jordan conveyed the Barreta grant to one of the attorneys.

Mrs. Durst never received any money from the sale of the land, and the two attorneys ended up with the land titles. Upon learning this, former Ranger Armstrong quickly suspected illegal activities. In 1883 the family set out to prove "that every act done by the [attorneys] was done in pursuance of a scheme and with intent to defraud the [Dursts]." Six years later, on March 26, 1889, after extensive legal dealings, the Barreta lands were legally and rightfully returned to the Durst and Armstrong families.[12]

During those years of frustration before vindication, Armstrong and brother-in-law James W. Durst, along with rancher Mifflin Kenedy, developed La Barreta as a productive ranching operation.[13] It was a genuine opportunity for John Armstrong to establish himself as a pioneer in the ranching business; whether he thought of himself as such is undetermined, but he would forever be known not only as the man who captured John Wesley Hardin but also as the man who helped drastically alter the environment of the Nueces Strip.[14]

There was considerable expense involved, as not only the funding for ranching had to be found but also the expense money for the trips to Mexico and elsewhere to obtain pertinent documents pertaining to land titles. A letter from Armstrong to Capt. Mifflin Kenedy is indicative of what Armstrong had to deal with. He had given attorney James B. Wells copies "of all the title papers in my possession together with all the data" he had gathered. Wells promised to go to

Corpus Christi to meet with Kenedy to work out an understanding of payment of expenses to be "incurred in going to Mexico in search of further evidence etc." Armstrong told Wells he was ready to go, and that Kenedy would no doubt contribute. "He [Wells] did not write me from Corpus as he promised," complained Armstrong. Then on July 9 Wells wired him from Brownsville that he had just reached there and would "write fully," but nine days later no letter had reached Armstrong. "I know how *unreliable* he had always been on the letter question," continues Armstrong, "but thought in *this* case he would keep his promise" (original emphasis). In the meantime, Armstrong had received other important information from Mexico which convinced him "that I can get all we will need, & acting upon this I have paid Mr. [Bethel] Coopwood $1000.00/100 to go with me & to aid also in the suit." Coopwood read Spanish and understood Spanish law "better than any other man in the state."[15] Armstrong believed that he and Kenedy would go to Victoria, Tamaulipas, and probably on to San Luis Potosí, San Luis Potosí, in Mexico. He suggested that, if Kenedy felt "disposed to aid me in this matter, it would be very satisfactory to me to have you defray expenses of the trip," which he believed would be under one thousand dollars. "My time and work, & that of Mr. Coopwood, I think, would be a fair off set against the expenses. . . . Am confident we will win." Armstrong certainly felt Kenedy should pay part of the expenses but made it clear that he would make the trip regardless. "I hope . . . that you will have sufficient confidence in me to come up with your share of the expenses. . . . Please *wire* me upon receipt of this, your determination in the matter as we will be ready to start in a few days" (original emphasis).[16] Armstrong's strength of character and will comes out in every line of that document.

A follow-up letter written in mid-February 1884 to Captain Kenedy reveals Armstrong's growing optimism and discusses plans for fencing land on La Barreta. He had just returned from Brownsville and met with Mr. Wells to discuss plans. "He like all the other attorneys to whom the matter has been submitted thinks our title good. We will manage to cancel the certificates at next term of court unless they are removed before that time." At this time Kenedy had acquired eleven of the twenty-five leagues of land, and Armstrong and the Dursts owned the other fourteen leagues. Armstrong now suggested making a partition "by mutual agreement so as [to] enable us to join in partition fence and each have the part he owns in a separate pasture, so as to stock as he may deem proper."[17] Fencing large tracts of land was just one example of Armstrong's ability to pioneer in a harsh land.

A near tragedy marred the trip home from Brownsville, as some twelve miles above Edinburgh, "the horses ran away with the stage & made a general *smashup*. Mrs. D. received a very painful injury of the knee & also part of the wrist, from which she is now confined to her bed. No bones broken, but severely sprained both limbs." Armstrong did not believe the injury would be permanent and hoped "she [would] soon recover." They arrived home on February 14.[18] With such real concerns for Mrs. Durst's well-being, a child to rear, and another on the way, along with the stress of obtaining a full title to La Barreta, Armstrong may well have occasionally mused about how much easier life had been as a Texas Ranger.

A further trace of optimism is evident in another letter written a month later, also to Captain Kenedy. Armstrong first expresses his feeling toward Attorney Coopwood, who was derelict in his duties in getting some important documents in Spanish translated. He writes, "He, like all other attys[.] with whom I have dealt is often 'guilty of willful neglect of duty.'" But more important, "There is sufficient evidence in the documents in my possession to establish beyond question the position of 'La Barreta' as it has always been supposed, claimed and understood to be—that is, adjoining the Carricitas on the North, Laguna Madre on the East and running back for quantity."[19]

As for the division of land, Armstrong had no objection "to you taking your interest as indicated" and felt that James W. Durst would have no objections either. Durst had been to see Armstrong some two weeks before and says in the letter that "he would like to keep the ranch & the *rincon* [corner] where he now has his cattle & for that reason hoped you would select the western portion." In the same letter Armstrong tells his friend that Mrs. Durst is able to walk "a little with crutches & we trust will soon be well again."

A strange twist is raised in a letter three months later. Now it appears that Armstrong was considering selling the land and establishing himself permanently in Austin, giving up his ranching interests. To Captain Kenedy he writes that he is "aware of the large bodies of good grazing in the state [that] have gone into hands that will hold them . . . but I have made up my mind to sell our interest. We can never live down there." He adds that his seventy-two-year-old mother is living with them.[20] Mrs. Durst, Mollie's mother, also lived with Armstrong, his wife and two children, and "a precinct to hear from" soon.[21] Further, he notes in the letter that he bought a beautiful home the previous fall on which he still owes $2,000. The children "will soon be ready to start to school; we want to live here & are happy in our home & its surroundings &c, all these things with the worry of taxes, the need of money to fence & stock our land, to say nothing of the worry & loss of time & money in hunting up our title &c, I say,

The Armstrong residence at 2610 Whitis Avenue, Austin. *Courtesy Austin History Center,* Pich 02939

taking every thing into consideration, we think it is not a bad idea to sell, & will be Satisfied with the price we can get." [22]

Fortunately, the period of angst was not long-lived and the land was not sold. Perhaps the fact that Armstrong had put his thoughts down on paper was enough to resolve the feeling of being overwhelmed by family responsibilities. Sometime during 1885 Mr. and Mrs. Armstrong and their three children, Maria Josephine, Jamie Durst, and John Barclay, along with Mollie's mother, moved to the seemingly barren and isolated La Barreta land. Four more children would be born to them after moving to South Texas. [23]

The happy family was welcomed by James W. Durst, Mollie's brother, who had been living on the land since around 1883 to validate the Durst-Armstrong claim. Daughter Maria Josephine later recalled arriving at their new home. "We arrived about at sundown and Uncle Jim walked out to meet us and welcome us," she recalled. She described the house as a "very primitive, temporary home on the ranch." It had but one story with a front porch and a small extension in the back, which was connected and used as a dining room and kitchen. "A heavy tarpolin canvas was used to divide the main house into two rooms." It was there that the family lived while Armstrong and old don Fermín, a "good Mexican carpenter," built the ranch house "about two miles south where there was good water." [24]

Armstrong and don Fermín completed the building in eighteen months. It was situated on a beautiful motte of live oak trees and became known as the Chicago Ranch, although how this name came to be chosen is unknown. Maria Josephine, the firstborn, married Andrew Stewart of New Orleans in 1906, and recalled that their home "consisted of [the] main house and a group of *jarcals* [*jacales*] grouped around a big tree making almost a patio and consisting of dining room and kitchen together, and two other cottages, then the Mexican quarters further away by the cattle pens." The main house and small cottages had thatched roofs made by Mexicans of sacahuiste grass. [25]

The Chicago Ranch was moved some ten miles from the original site in 1893. Josephine Stewart is our authority on its appearance at the time, writing that the "new ranch home was built facing south with wide porches front and back and extensions at each [end] and going backward forming a patio enclosed on three sides." Ultimately, an "enormous grape vine was trained upon a telegraph pole in the center of the court and it formed a tent like court of shade as well as looking very picturesque . . . along the back porch." By then Armstrong had enclosed a bathroom for every room. Today John B. Armstrong's descendants reside in this residence. [26]

The Chicago Ranch house, built by John Armstrong and don Fermín in 1895. *Courtesy Tobin Armstrong*

In 1983 a state historical marker was unveiled on the Armstrong Ranch, some twenty miles south of the Kenedy County seat of Sarita. The marker text reads:

In 1852 James H. Durst, son of a leading Nacogdoches, Texas family, purchased 83,219 acres of land here, part of "La Barreta" Spanish Land Grant. In 1878 Mary Helena "Mollie" Durst, daughter of James and Mary Josephine Atwood Durst, married the noted Texas Ranger John Barkley Armstrong. Armstrong had served with Captain Leander McNelly and played a major role in bringing law and order to South Texas. He participated in the arrest of King Fisher and gained national fame for his capture of the notorious Texas outlaw John Wesley Hardin. Armstrong moved his family to the ranch house he built here. Their close friends and neighbors were the families of Captain Richard King and Captain Mifflin Kenedy. The ranch was an important site in the area; General Zachary Taylor had camped here prior to the Mexican War and for many years the ranch served as a stop on the stage route between Corpus Christi and Brownsville. Under Armstrong's guidance, the Armstrong Ranch became one of the legendary cattle ranches of Texas. His descendants have continued the tradition of family enterprise here until the present time.

The most frequently published portrait of John B. Armstrong, probably taken in the 1890s. *Courtesy Western History Collections, University of Oklahoma Library, Norman*

CHAPTER NINE

Rancher Amid the Rails

[John B. Armstrong was] a pioneer
who blazed the path for modern civilization.

—*SAN ANTONIO DAILY EXPRESS*, MAY 2, 1913

O N APRIL 3, 1888, John B. Armstrong mustered in as a private with the Brownsville Rifles and was promoted to the rank of major and division quartermaster on January 2, 1893. Eight days later he was promoted to the rank of major and assistant inspector general. From May 1, 1895, to July 13, 1900, Armstrong was lieutenant colonel and assistant chief of ordnance, but he always preferred to be called Major Armstrong. During the Spanish-American War he assumed command of a Texas volunteer regiment and accepted the rank of major given him by Gov. Charles A. Culberson. Armstrong never cared to talk about the Spanish-American War, because, as he said, he never had the chance for "real service" when there was fighting, but he would have fought had the chance been afforded him.[1] Curiously, some historians believe that the rank of major came from his service in the Texas Rangers.

Not surprisingly, after resigning from Ranger service at the end of 1878, Armstrong reenlisted, in the Frontier Battalion, Company D, on June 5, 1888, as a "Special Ranger." Although he may have thought of rejoining the Rangers on his own, Robert J. Kleberg took the initiative to recommend him and several others to the force. Kleberg corresponded with Adjutant General King on May 29, 1888, referring first to the "large scope" of country along the boundary of Nueces, Cameron, and Hidalgo counties, where the three joined, which was "so remote" from the county seats of those counties that it was impossible for

A major and four Texas Ranger captains, circa 1900. Standing from left: John A. Brooks, John H. Rogers, unidentified. Seated from left: Lamartine Pemberton "Lam" Sieker, John B. Armstrong, William J. McDonald. The unidentified standing figure at right may be Adj. Gen. W. H. Mabry or a man named Waite; it is definitely not J. B. Gillett, as has sometimes been claimed. *Courtesy Haley Memorial Library and History Center, Midland, Texas*

the respective sheriffs to have any real influence on "violators of the law." The remoteness allowed lawbreakers to slip easily from one county to another, but the sheriff's authority ended at the county line. Each county found itself with insufficient funds to hire deputies who could patrol the areas "in such remote and unsettled precincts." [2]

Kleberg suggests in the letter that Adjutant General King present the names of "some of the leading citizens of that part of the Country" as State Rangers. They would not receive any pay and thus would not cost the state or counties anything; they would "supply themselves & can protect themselves & property." This amounted to giving various "leading citizens" the legal right to become hired guns. Kleberg provided Adjutant General King with the names of Armstrong; John G. Kenedy, the son of Mifflin Kenedy; former Texas Ranger E. R. Jenson; Armstrong's brother-in-law, James Durst; and Paulin S. Coy, who also was a former Texas Ranger. [3]

Special Rangers, as noted above, received no payment from the state, had to apply for the position on the approved form, were assigned to one of the Frontier Battalion companies, and the last day of each month were to report by letter to the Adjutant General's Office the number of arrests made and any other relevant activities.[4] Armstrong described himself as thirty-eight years of age, brown hair, five feet eleven and one-half inches tall, of light complexion, and born at Readyville, Tennessee. He was enlisted by Adjutant General King at Corpus Christi. During the early 1870s, while riding with McNelly and Hall, no such descriptive list was required. Now the document was to be carried by every Ranger "and will be exhibited as a warrant of his authority as such, when called upon, and must be surrendered to his Company Commander when discharged."[5]

Robert J. Kleberg, circa 1875. *Courtesy King Ranch Archives, King Ranch, Inc., Kingsville, Texas*

Armstrong was assigned to Capt. Frank Jones's Company D. Jones had an impressive record of service to the state. The son of Judge William E. Jones of Austin, Frank first enlisted in Lt. Ira Long's Company A in September 1875, when he was nineteen years of age. He served almost continually in the service, with Long, then with Capt. Pat Dolan in Company F, then in Company D under Capt. D. W. Roberts, then again under Capt. L. P. Sieker in Company D. By 1884 he was a lieutenant; on May 1, 1886, he was promoted to captain. This shining star was shot and killed by bandits in a battle near the Rio Grande on June 30, 1893.[6] When Armstrong was assigned to his company the camp was in Duval County. The county seat — San Diego — was nearly seventy miles from Armstrong's ranch headquarters. Others who were assigned to Jones's company at this time were John Chenneville, Paulin S. Coy, G. B. Greer, W. J. Greer, A. M. "Gus" Gildea, E. R. Jenson, John G. Kenedy, Sam R. Pickett, J. E. Van Riper, and Ernest Rogers.[7] Most men who became Special Rangers had previous law enforcement experience. Chenneville had been a policeman in Austin; Coy, Greer, Jenson, Kenedy, Pickett, and Van Riper had all served as regular Texas Rangers.

Van Riper reported his occupation as a deputy U.S. marshal, and it is possible that from this it was commonly believed that Special Rangers held the same authority as deputy U.S. marshals. Nothing in the available record, however, indicates that Armstrong ever worked in that capacity. *The Handbook of Texas* provides a brief paragraph on his life and career and states that, after retiring from Ranger service, he "became a United States marshal." This error is maintained in a larger entry for him in *The New Handbook of Texas*. If this is indeed true there should be verification in the form of a commission or pay voucher—some contemporary form of documentation that he served as a deputy U.S. marshal. I queried the National Archives and received this reply: "An examination of the records of the Department of Justice in our custody fail to reveal any mention of John Barclay [sic] Armstrong serving as a United States marshal or deputy marshal or any documents concerning his possible appointment. A further examination of the *Official Register of the United States Government* failed to show that Armstrong worked for the Justice Department in any capacity between 1877 and 1882." The source of this confusion may have begun with the 1916 publication of *A History of Texas and Texans.* There Frank W. Johnson writes that Armstrong's career, "both as a ranger and a deputy United States marshal, brought him, in fact, in conflict with nearly all the famous criminal characters of Texas, the list including Sam Bass, Ben Thompson, Alfred Aylee [Allee], and numerous others. After retiring from the ranger service he was appointed deputy United States marshal and served in that capacity until the time of his death." [8]

By November the company was stationed in Rio Grande City, county seat of Starr County, and Armstrong was still listed as a Special. The February 1889 muster roll still carries his name, but W. J. Greer is no longer listed; G. J. Reynolds has taken his place. [9] Now the company was stationed at Uvalde and was still stationed there in May, but now with C. L. Broome and J. W. Mathers having joined. [10] Armstrong's name is still carried through August and again in February 1890, when the company was stationed at Cotulla, La Salle County. In August and still in November 1890, when the company was in Brewster County, Armstrong's name is on the roll, but for the last time.

Each Special Ranger had to have a captain to whom to report. What actual duties Armstrong performed as a Special Ranger are not recorded, as, apparently, any written reports he forwarded to Capt. Frank Jones have not survived. He may have met Jones only once, when he joined Company D. None of Jones's monthly returns carry Armstrong's name; only the muster rolls prove he did serve.

Armstrong was also required to take an oath when reenlisting. His is dated July 5, 1888, and was sworn to and subscribed before notary public T. P. Rivers of Nueces County: "I do solemnly swear (or affirm) that I will faithfully and impartially discharge and perform all the duties incumbent upon me as a member of the Frontier Battalion until August 31, 1889, unless sooner discharged, according to the best of my skill and ability, and agreeably to the Constitution and laws of the United States and of this State. So help me God."

One incident that has been recorded involved Armstrong only indirectly.

On May 16, 1902, on Rancho El Sauz, a component of the King Ranch, Anderson Y. Baker, a former Ranger in Capt. J. A. Brooks's Company A, came upon a rustler, Ramón Cerda, branding a stolen calf. Cerda resisted Baker's demand to surrender, instead firing his weapon. Baker's aim was superior and Cerda was killed. The Cerda family attempted to create sympathy for the dead rustler and even put a bounty on Baker's head. Naturally, in the ensuing troubles between the Cerda family and authorities Armstrong, as well as the King Ranch, supported Baker's action. It was only chance that it had been Baker and not Armstrong who came across the man illegally branding the calf.[11]

Because some Rangers around the turn of the twentieth century were not investigated thoroughly before being selected, many Mexicans did receive summary "justice" from men who should never have been sworn in. No writer has ever accused Armstrong of this brutal type of justice. In fact, any known victims of Armstrong's were killed when in the act of shooting at him or his men. The fight at Espantosa Lake resulted in the deaths of several suspected thieves, but they were all Anglos—none were Mexicans—and we do not know whether any were in fact killed by Armstrong's bullets. The gunfight resulting in Hardin's capture and the death of Jim Mann involved only Anglos. Did Armstrong kill anyone during the invasion of Mexico in November of 1875? There is no way of knowing.

Although some writers have attempted to create the image of the Texas Ranger as someone who practically walked on water, some Rangers were indeed of poor quality. McNelly produced many good men who came to him in their late teens or early twenties, but his muster rolls also reveal that he was not afraid to discharge those who did not come up to his standards. Every captain probably had to rid his company of undesirable recruits at times. Robert M. Utley perhaps summarizes it best: although "McNelly and his men unjustifiably abused Mexicans," by the aftermath of the Civil War the "outrage" over such incidents as the Alamo, Goliad, "and other Mexican offenses of the revolution and the republic subsided." Further, the Rangers "embodied the attitudes of Anglo Texans and also of Mexicans" toward Indians. When engaged in combat with Indians the Rangers "paid no more heed to humanitarian principles than their foe."[12] It was a violent time in a violent country filled with violent men. All who attempted to live and survive in the Nueces Strip had to be able to defend themselves and their possessions, or else they would not prosper, let alone survive.

As the Travis Rifles had done years before—provide drill competitions for the public—so did the Brownsville Rifles now. In June of 1889 a huge competition was to be conducted in Galveston, with teams competing from not only

Texas cities but out of state as well. The reason was the celebration of Galveston's semicentennial, which would take place June 4–15. Broadsides were being
circulated as early as March. The "official circular and prospectus" promised
prize money aggregating $20,500. Circular No. 1 is dated March 16 and prom-
ises a $3,000 prize for the winner of the "Grand Interstate Infantry Drill," with
$1,000 and $500 for second and third prizes, respectively. State infantry companies were eligible for eight prizes, ranging from $1,000 for first to $200 for
eighth prize. The company which ranked highest in camp discipline would receive a handsome flag valued at $250. Not only were there to be drill competi-
tions but for everyone's enjoyment there was to be a "grand sham battle, engag-
ing all the volunteer organizations and the whole force of regular troops on the
ground." There would also be a "grand military street parade."[13]

Eligible Texas infantry companies were those that had not previously won
a prize in an interstate drill competition with other interstate companies. The
team had to be composed of no fewer than sixteen men, two guides, and three
commissioned officers. The first prize went to the Brenham Light Guards. Apparently, the competition was stiffer than anticipated as the Brownsville Rifles
took ninth place out of nine teams. But things went better in the "Texas Maiden
Infantry Companies" drill, as out of five entries, the Victoria Rifles took first
with a prize of $500; second prize went to the Brownsville Rifles, who won a
purse of $400.

Armstrong's name does not appear among the prizewinners, but the Brownsville Rifles took first place in the category of "best camp inspection, being most
quiet and orderly in camp, promptness in obeying orders, having best policed
quarters, whose detail in promptness in reporting for guard mounting and dress
parade." The prize was the handsome flag.[14]

Two years later, Armstrong resigned his position. His letter to Adj. Gen.
W. H. Mabry in Austin, dated March 6, 1891, and written on the letterhead
of the Corpus Christi and South American Railway Company (of which he was
general manager), reads as follows: "In compliance with General order No 2, of
21st last month, I have the honor to return herewith my appointment as 'Special Ranger,' issued by Adjt. King, 15th June 1888."[15]

But on April 28, 1891, Armstrong again swore an oath to support the Constitution of the United States, to bear true faith and allegiance to the State of
Texas and the Constitution thereof, and to further "faithfully observe and obey
all laws and regulations for the government of the Volunteer Guards of this
State, and the orders of all officers elected or appointed over me, for the period
of three years." This document was sworn and subscribed before notary public

James B. Mitchell at Corpus Christi. Armstrong
again gave as his occupation manager of the Corpus Christi and South American Railway Company. He was enlisted by Gen. A. S. Roberts, also
a member of the Travis Rifles back in the 1870s
in Austin.[16]

Little is known of Armstrong's official activities during this period. On July 14, 1892, a "General Court Martial" was appointed to convene at
division headquarters at Camp Mabry in Austin. The detail for the court consisted of Col. A.
Faulkner; Lt. Col. W. Von Rosenberg Jr.; Maj.
J. B. Armstrong, division quartermaster; Maj. A.
Harrison, artillery battalion; Capt. J. W. Walies of
the Rushford Rangers; and Capt. J. M. Goggan
of Eagle Pass, Retired. Lt. Col. John Duval was
judge advocate.[17]

Armstrong as a Special Ranger, circa 1889.
Courtesy Tobin Armstrong

Armstrong had had experience with these
types of administrative duties. Special Order
No. 182 had placed him at the head of a detail
of the Board of Inspectors to meet at Corpus
Christi on July 1, "or as soon thereafter as practicable" to pass upon and recommend the disposition of certain forage caps, blouses, trousers, chevrons, and
pant stripes for which Capt. C. B. McCampbell of the Texas Volunteer Guards
was responsible. Others making up the board were Maj. R. W. Stayton, assistant
judge, and 2nd Lt. D. G. Brewster of Corpus Christi.[18]

His oath was affirmed a year later, on April 8, 1893, but now before F. W. Sea-
bury, notary public for Cameron County, at Rancho Chicago, and Armstrong's
occupation was given as "Ranchero."[19] The oath was administered again two
years later, on June 26, 1895, now at Alice, Jim Wells County, before F. B. Sayer,
notary public. Armstrong's occupation this time was given as "Stock Raiser."[20]
A final oath in his file is dated at Austin, January 12, 1900, sworn to before
Samuel D. DeCordova, Travis County notary public.[21] The description now
shows him as being fifty years old, as having been born near Murfreesboro, Tennessee, and as a "Cattle Raiser." It also shows his military service as about sixteen years in the Texas Volunteer Guard and Texas Rangers. Additional information provided includes his post office and telegraphic address as Austin and
Alice, his Austin residence as 2610 Whitis Avenue, and his Austin telephone

number as 641.[22] One document, which may have been obtained in order to reenlist, is signed by J. F. Cummings, formerly captain of the Brownsville Rifles. In this document Cummings certifies that Armstrong "became a member of the Brownsville Rifles on April 3, 1888 and that he remained a member of said company until he was appointed and commissioned Division Quartermaster with the rank of Major on Jan. 2ᵈ, 1891."[23]

It was during this period that Armstrong the ranger and rancher established himself as Armstrong the pioneer as well. When he first entered the Nueces Strip back in the mid-1870s it was frontier, peopled by a very few courageous ranchers, such as Richard King and Mifflin Kenedy. Their huge landholdings were unfenced, their cattle roamed free but were susceptible to raiders from south of the Rio Grande as well as Anglo outlaws ravaging herds far from their owners. John B. Armstrong, experienced in dealing with strong men, real estate, cattle and ranching in general, now became more determined than ever to develop his land into full production with him as the operator. The land was fenced with materials shipped in from New Orleans. Ranching was pursued "on a major scale," and Armstrong was determined to "provide water for livestock, whether it be for the indigenous or feral animals or their own cattle." Armstrong and the operators of the King and Kenedy ranches "developed equipment for fencing for domestic cattle. They cross bred their domestic stock with wild cattle, bred shorthorn cattle, Hereford, Angus. The shorthorn or Durham was the third most popular breed. The earlier feral cattle were also known as 'Spanish cattle.'" During this period of ranching history and pioneering, essentially the decades from the early 1880s through the remainder of Armstrong's life, these developments proved to be effective for the ranching industry: "They discovered artesian wells, water in the Goliad sand, nine hundred feet down. They drilled the ranch on a systematic basis at regular uniform distances."[24]

And at the same time these pioneer ranchers were working with the railroad in an effort to provide a market as well as transportation for their ranching operation. Before the railroad was established, cattle herds had to be driven to market in either Corpus Christi or Brownsville. Armstrong may never have driven longhorns up the cattle trails to the Kansas markets, such as Abilene, Wichita, or Dodge City, but he certainly had experienced driving cattle in the Nueces Strip. He knew there had to be a better way, and he was among the driving forces to bring the railroad to South Texas.

The greatest tragedy of Armstrong's life occurred in December 1897. On that day his beloved wife, Mollie, took a break from practicing the piano and went outside to the front of the house and began playing with a Newfoundland dog

The Armstrong family, circa 1890. Top row from left: John B., holding Charles on his lap; Julia; Josephine; "Granny" Durst; Jamie; Mollie holding Elliott "Tim." Seated in front is John. *Courtesy Tobin Armstrong*

that belonged to their youngest daughter, Julia. The dog scratched her forearm with a tooth, causing a minute abrasion. Mollie washed her arm and thought no more of it. It was about the time that they had planned to move back to Austin so the children could attend good schools and she could be once again surrounded by the friends of her youth. But ten days later, Mollie became ill. The dog had given her more than a scratch as it was suffering from rabies. On Christmas Day, 1897, Mollie Durst Armstrong's agonies ceased and she passed on. When it was apparent that she was gravely ill efforts were made to notify Armstrong. He could not be reached for several days, so the responsibility of caring for her on what became her deathbed fell to daughters Josephine and Jamie. Ironically, the two girls had left their New Orleans college, Newcomb, to attend the University of Texas, because of a yellow fever scare in that Louisiana city.[25]

Armstrong was on his way to Brownsville when a telephone message caught up with him from Robert Kleberg, the best man at their wedding, advising him of Mollie's illness. He was eighty miles from Santa Gertrudis, the King Ranch, and when he arrived there he learned of her death. Kleberg and Armstrong left immediately for Austin, arriving there early in the evening of Monday, December 27. The funeral took place the next day at eleven o'clock, at St. David's Church, the same church where they had been married almost twenty years before. Mollie Armstrong was buried in Oakwood Cemetery in Austin.[26]

Dealing with life's natural tragedies did not break Armstrong's will to survive and contribute to society. As important as his contributions may have been to the Texas Volunteer Guard, certainly more beneficial was his part in getting railway service for South Texas, which brought not only increased population but access to the towns and additional markets. He approached officials of the monopolistic Southern Pacific to extend their rails into the area south of Corpus Christi. At first he was encouraged that his efforts would produce results, but after a time railroad officials decided the cost would be prohibitive. This angered not only Armstrong but many of the leading citizens of Nueces, Hidalgo and Cameron counties, including Mrs. Henrietta King, Richard King's widow; Robert Driscoll; and John G. Kenedy.[27]

Uriah Lott had established a solid reputation as a successful promoter, having proved his worth in improving the Corpus Christi ship channel. One early accomplishment was building the Corpus Christi, San Diego and Rio Grande Narrow Gauge Railroad from Corpus Christi to Laredo. In late October 1902 Lott was on a "confidential railroad mission through Texas" looking over a proposed line. His trip was successful, and on January 12, 1903, a charter was issued to a corporation organized for the purpose of "constructing, owning, maintaining and operating a railroad from Sinton in San Patricio County, south across Nueces, Hidalgo and Cameron counties 160 miles to the town of Brownsville." This line was to be known as the St. Louis, Brownsville & Mexico Railway. The incorporators were Robert J. Kleberg, A. E. Spohn, Robert Driscoll Sr., Uriah Lott, Richard King II, John G. Kenedy, James B. Wells, Francisco Ytúrria, Thomas Carson, Robert Driscoll Jr., E. H. Caldwell, George F. Evans, Caesar Kleberg, John J. Welder, and John B. Armstrong.[28]

The cooperation Armstrong and the others got from Uriah Lott was not mirrored in their experience with the operators of the Southern Pacific. Besides Mrs. King, Driscoll, Kenedy, and Armstrong, in particular, donated large tracts of land required by railroad building contractors. They all finally experienced success when, on Independence Day, 1904, a "bunting-draped excursion train northbound from Brownsville to Corpus Christi" made the maiden run. The first train left Brownsville at 6:30 AM and then the telegrams poured in to President Lott, congratulating him on the completion of "one of the most substantial railroads in the State." The whole town was jubilant, anticipating "the great outside world. . . . All the public offices and banks were closed and the buildings were decorated in honor of the first train arriving in Brownsville the night before at 7:20 o'clock." Some seventy-five passengers were on the train, and they were greeted with a band and escorted to downtown "by a constant display of fireworks and a grand hurrah for Colonel Lott." [29]

The development of the railroad in South Texas saw the birth of numerous towns. Robstown, named after Robert Driscoll Jr., began as a material yard where railroad ties soon smothered the mesquite grass. The town of Driscoll soon followed, then Bishop, first named Julia after a daughter of Robert Driscoll Jr., then changed to Bishop. This was followed by Caesar, named for Caesar Kleberg. Kingsville, first surveyed in May of 1903, was aptly named for Richard King. Then came Ricardo, named after Richard King II, followed by Riviera, first named Spohn; then Sarita, today the county seat of Kenedy County, named after John G. Kenedy's daughter. The towns of Mifflin, Turcotte, and Armstrong followed on the line.

The town of Armstrong was first named Katherine, after daughter Julia Katherine, then rechristened Armstrong due to the frequent confusion with the similarly named town of Caterina in southern Dimmit County. The track reached this point on March 4, 1904.[30]

But for all the joy and exuberance of the railway and the opportunities it offered people and the wealth it brought to the Nueces Strip, there were still tragedies in Armstrong's life. The pain of Mrs. Armstrong's death of course could not be equaled. But then on May 6, 1905, the Armstrongs' first son, John Barclay, born on July 7, 1884, lost his life. The young man was working a cattle roundup when several cattle became embroiled in a fight into which John's horse could not avoid entering. When he was thrown young Armstrong's neck was broken and he died instantly, with his father watching helplessly, unable to do anything to save his life.[31] Yet another tragedy struck the family with the death of Mary Josephine Atwood "Granny" Durst on September 10, 1912, at the home of her sister, Mrs. August E. Palm, in Austin. Born August 21, 1830 in Bolivar, Tennessee, at the age of eighty-two years she was certainly prepared for death, having been in failing health for almost a year. As the *Austin Daily Statesman* expresses it, "although medical science and loving care prolonged her life longer than had been thought possible in the beginning of her illness, her advanced age and the seriousness of her malady were odds too great for earthly ministration." [32] She had come to Texas with her parents, William and Mary Neely Atwood, in 1839. She had married James H. Durst on January 11, 1854, and with her husband lived in the Lower Rio Grande country, where Major Durst was collector of revenue.

After Durst's early death she moved to Austin with her two children, Mary Helena "Mollie" and James. At the time of his death Major Durst left a large estate including La Barreta. The many slaves he owned were freed after the Civil War. When Mollie died in 1897, Mrs. Durst "took entire charge of the house and younger members of the family and ministered to their wants and interests with all the tenderness of their own mother, a fact which is singly testified by the devotion which they have each shown her, especially in her last months and weeks of illness." The funeral was held at St. David's Church in Austin. Along with her name and dates of birth and death, her tombstone in Oakwood Cemetery

Major Armstrong with sons John and Charley, circa 1896. *Courtesy Tobin Armstrong*

reads, "And they shall be mine saith the / Lord of hosts, in that day when / I make up my jewels." [33]

Jamie Durst Armstrong, born January 5, 1881, married John M. Bennett of San Antonio on April 26, 1905, at St. David's Church in Austin. The Rev. Dr. T. Booth Lee officiated. Her sister Maria Josephine was her attendant while Bennett was attended by George Maverick. A reception followed at the Driskill Hotel. Following the reception the couple left on the International and Great Northern train for New York from where they would sail for Europe to visit during that summer. They then made their home in San Antonio. She died there on April 23, 1963.

On April 26, 1913, a Saturday, Major Armstrong felt the effects of an unidentified illness which took his life on Thursday May 1, at 2:55 PM. Dr. M. E. Miles, who attended Armstrong from April 26 to the day he died, wrote on the death certificate that death was caused by "Uremia due to chronic intestinal nephritis." [34] The news spread from Kingsville that the old Ranger had passed away "peacefully and without a struggle." [35] Daughters Josephine and Julia were at his bedside when he breathed his last, as were sons Charles M. and Thomas. Daughter Josephine was unable to be there because she was confined at home with illness. A private car of the Brownsville Road, of which Armstrong was a director, transported his remains from Kingsville to San Antonio, where the car was transferred to the Missouri, Kansas & Texas tracks. The car arrived in Austin at noon on Friday, May 2. Accompanying the remains were daughters Jamie and Julia, sons Charles and Tom, Mr. and Mrs. Robert Kleberg, Miss Henrietta Kleberg, John G. Kenedy, Dr. Shelton, Judge Claude Pollard, and Jeff McLemore. [36]

At St. David's Episcopal Church, at 5:00 PM on May 2, the coffin was laid on a bank of flowers and was surrounded by a large number of mourners. The Rev. Milton R. Worsham conducted the religious services. A sextet composed of Misses Louise Pfaellin and Jessie Smith, Mrs. Eugene Haynie, Mrs. M. Mitchell, W. H. Stacey, and C. M. Calloway sang "Lead, Kindly Light," "Abide with Me," and "O, Mother Dear, Jerusalem." Armstrong was buried in Oakwood Cemetery in Austin. [37] Active pallbearers were Jeff McLemore of Houston, Ike Champion, Ed J. Byrne,

John B. Armstrong shortly before his death in 1913. *Courtesy Tobin Armstrong*

Grave marker of John B. Armstrong and
members of his family, Oakwood Cemetery,
Austin, Texas, 2003

Lee Borden, T. J. Lollis, and Dr. H. L. Hilgartner. Honorary pallbearers were former governor Joseph D. Sayers, Judge F. A. Williams, Judge T. S. Maxey, E. M. Scarbrough, C. D. Johns, and Col. W. C. Gaines.[38]

A twenty foot–tall marker, in the form of a Celtic cross, marks his grave. At the top is inscribed, "I am the resurrection and the life" and "Glory to God in the Highest." Also inscribed several times among the decorative leaves is "IHS"—In His Service—and the following:

> JOHN B. ARMSTRONG
> 1850–1913
> Reverent toward God
> A loving husband
> A devoted father
> A faithful friend.

Henry Hutchins, adjutant general, on May 2 issued General Orders No. 22:

It is with sorrow, the Commander-in-Chief, announces the death of Lieutenant Colonel John B. Armstrong, retired, which occurred at Kingsville, Texas, May 1, 1913.

The service of John B. Armstrong as enlisted man and officer in the Ranger Force was brilliant, and in later years his services in the various staff departments of the Texas Volunteer Guard were of great value.

The order shows Armstrong's record: as a private under Capt. L. H. McNelly, May, 1875; as a sergeant, July 1875; second lieutenant, April 1, 1877; first lieutenant, December 1, 1877 to January 1, 1879. In the Texas Volunteer Guard he joined as a private in Austin, serving from 1873 to May 1875; then as a private in the Brownsville Rifles, April 3, 1888; then as major and division quartermaster, January 2, 1891; then as major and assistant inspector general, January 10, 1893; and finally as lieutenant colonel and assistant chief of ordnance, May 1, 1895, to July 13, 1900. The general order was issued by Gov. Oscar B. Colquitt, commander-in-chief.[39]

Of the various surviving newspaper accounts the *San Antonio Express* article reminds readers of the most notable feat of Armstrong's career as a Ranger. It points out that he "was a Ranger captain from 1872 to 1878 [*sic*]. The performance that perhaps gave him the greatest fame as a Ranger captain was his arrest of John Wesley Hardin." The article continues, telling readers that San Antonio knew the man "as a pioneer who helped blaze the path for modern civilization, as a man who did much to root out lawlessness, as a rancher and cattle baron and then, who in recent years lent his efforts in turning the sunbaked plains into verdant fields." And the newspaper does not overlook Armstrong's contribution to transportation, adding that he "offered much of his acreage for colonization purposes. He gave liberally toward the building of the St. Louis, Brownsville & Mexico Railway and has been closely associated with its policy of bringing homesteaders into that section of Texas."[40]

Julia Katherine Armstrong, born February 5, 1889, married Zeb Mayhew of Great Neck, Long Island, on December 31, 1913, eight months following her father's death. They were married in St. Mark's Episcopal Church in San Antonio. Mrs. Andrew Stewart of New Orleans was the matron of honor. Julia entered on the arm of her brother Charles and was met at the chancel by the bridegroom and his best man, Theodore Pratt of New York. The Rev. Philip Cook performed the marriage service. After a brief gathering of friends at the Bennett home the couple left to visit California, after which they made their home at Great Neck.[41] The couple divorced and Julia later married Allard Kaufmann in New Orleans on Mardi Gras Day, 1936.[42]

Charles Mitchell Armstrong was born November 8, 1886, and married Lucie Tobin Carr, great-granddaughter of John William Smith, last messenger of the Alamo. They too were married at St. Mark's Episcopal Church in San Antonio, on January 23, 1918, with Bishop J. S. Foster officiating. The bride was attended by her sisters, Mrs. Warren P. Colvert and Mrs. W. Scott Schreiner, and was given in marriage by her father, J. M. Carr. Caesar Kleberg of Kingsville was the best man. After the ceremony an informal gathering of friends and family was held at the home of the bride's parents, and later they left for New Orleans. On their return they made their home on the Armstrong Ranch. Charles died on September 13, 1941, when his car overturned near Kingsville. The *Austin Statesman* describes him as a "prominent rancher and polo enthusiast."[43] He was buried in the Armstrong family plot in Oakwood Cemetery.

Elliott Ropes "Tim" Armstrong was born October 9, 1890, and died May 31, 1898, of diphtheria. He too is buried in Oakwood Cemetery. Thomas Reeves Armstrong was born September 5, 1892. He married Henrietta Kleberg Larkin on June 4, 1949, in the Church of the Holy Trinity in New York City. He died March 3, 1986. Firstborn Maria Josephine, born April 5, 1879, died on October 3, 1972. Jamie Durst, born January 5, 1881, died on April 21, 1963. Julia Katherine, born February 5, 1889, lived the longest of the Armstrong children, dying on December 26, 1991. John Mirza Bennett Jr., son of Jamie Armstrong and John Bennett, shared several memories with me in the form of attachments to his correspondence. One is entitled "Early Memories of Armstrong Ranch." He recalls that, when the Gulf Coast Railroad was built through the town of Armstrong, the station was established about two miles from the ranch house. This distance was covered by two horses pulling a carriage. In about 1911, when John was three years old, he and his mother and older sister were on a visit to Armstrong. Major Armstrong "met us at the station with a carriage and we started the trip across the pasture to the ranch house. For some reason, I started making childish noises in the form of a cry, which was probably based on little cause," he recalls. His grandfather told him to "hush-up" but apparently the order was not obeyed. "At this point he instructed the driver to stop the carriage and I was removed from the vehicle and left alone on the prairie while the carriage with my mother, grandfather and sister proceeded on to the ranch house. I can well imagine the feelings of my mother when her 'dear' little three year old was left on the prairie to be devoured by coyote[s] and rattlesnakes." But in a very short time the carriage returned with only the driver and took little John on to the ranch house. "Apparently I learned the lesson that when my grandfather gave me an order it was to be obeyed without delay." [44]

Around the turn of the twentieth century Major Armstrong was on a train with daughters Josephine and Jamie when the following incident occurred. Mr. Bennett entitles this memoir "A Passenger Train Incident." Jamie and Josephine were "young ladies in their early twenties," he recalls, when they and Armstrong were on the train. Toward the front of the coach were two men who had "probably been drinking before boarding the train. They were using loud language" and Grandfather Armstrong "objected to some of the obscene words." He "politely walked forward and told the two gentlemen that there were ladies in the rear of the coach and that he would appreciate their being more careful with their language." The "old Ranger" returned to his seat where his daughters waited. But the obscene language continued, at which point Armstrong stood up a second time and walked toward the two men from behind and "grabbed each of them by the hair and bounced their heads together, turned around and walked calmly back to his seat. No further untoward language was heard." [45]

John B. Armstrong relaxing with two grandchildren, Elizabeth Bonneau Bennett and John Mirza
Bennett Jr., circa 1912. *Courtesy John M. Bennett*

As John B. Armstrong returned to his seat, was he not reminded that a quarter of a century before he had experienced a much more dangerous altercation in a railroad car in far-off Pensacola, Florida, an altercation which could have easily ended in death but instead resulted in the capture of the deadly John Wesley Hardin? One can only wonder.

Afterword

ELMER KELTON

JOHN BARKLEY ARMSTRONG served the Texas Rangers from 1875 through 1878, during what many would consider the classic frontier period of that organization. This was a time when the Rangers were able finally to set aside what had been their primary concern, the Indian problem, and concentrate on the outlawry which was increasing with rapid growth of the state's population.

Though it would be fifty years before the name would become official, the Rangers in their various forms had served Texas since land empresario Stephen F. Austin called them into service in 1835 to fight off hostile Indian incursions into his colonies. In those early years, volunteers in the so-called ranging companies were more or less a local militia, called upon for limited tours of duty, often only a few weeks or months, and perhaps only a single campaign. Sometimes they were paid, sometimes they were not, depending upon the state of the republic's and, later, the state's treasury at any given point. Texas in its early years was poor as a church mouse.

From the beginning, the Rangers gained a reputation for being willing to meet the enemy and engage him, whoever and wherever the enemy might be. Rangers helped Texian settlers escape Santa Anna's advancing Mexican army during what was known as the Runaway Scrape during the Texas revolution. Under the republic's president Mirabeau Lamar the once-weak force was considerably strengthened. Unlike predecessor Sam Houston, who favored diplomacy and friendship with the tribes, Lamar advocated an aggressive policy which he hoped would drive them out of Texas. He succeeded in doing so with the Cherokees. He had much less success with the Comanche and other frontier horseback tribes. Rangers were present at the Council House fight in San Antonio in 1840, when Comanche chiefs came in to parley about the release of white captives. The meeting exploded into a bloody street battle in which many Comanche and a number of white citizens were killed.

In retaliation, a huge Comanche invasion force pushed all the way to the Gulf Coast, attacking Victoria and destroying the seacoast town of Linnville. Rangers and volunteers waylaid the retreating Comanche at Plum Creek, near present-day Lockhart, and routed them despite being vastly outnumbered.

Texas Rangers established their fighting reputation during the Mexican War, when their old exploits sometimes amazed, sometimes dismayed the officers of the American army to which they were attached as scouts. They became known to their Mexican opponents as "*los diablos tejanos,*" the Texas devils.

During this antebellum period outlawry was a concern, but most Ranger effort was devoted to protecting settlers from Indians. Fast, briskly efficient, the Rangers could often strike retreating marauders while the more cumbersome federal army was still organizing for a start. As before, the size of the units expanded and contracted with the convolutions of the state treasury.

The Rangers went into one of their lowest periods during the Civil War because of funding shortages and a lack of manpower. Many fighting-age men went off to serve the Confederacy on distant battlefields far to the east. By the war's end, the Rangers had been reduced to a fragment, and the collapse of the Confederacy left them in complete limbo. During the Reconstruction years there was no Ranger force, not even on paper.

The Reconstruction governor, E. J. Davis, organized a force known as the State Police, modeled somewhat after the Rangers. His group concentrated much less on Indians than on outlawry and resistance to the occupying powers. Though it did more good than it has been generally credited with, it was also cursed with corruption and a number of "bad apples" who cruelly abused their power. It was disbanded even before the 1873 election which swept a defiant Davis out of office.

One of the first acts of his successor, Richard Coke, was to reestablish the Texas Rangers and form the now-famous Frontier Battalion. Unlike the loosely organized Ranger groups of earlier years, this one was tightly knit like an efficient military unit, well equipped to fight the lawlessness that prevailed in the wake of the war and the turbulent Reconstruction years.

It was into this new and highly mobile force that John B. Armstrong stepped in 1875. He served concurrently with such legendary officers as Leander H. McNelly, John B. Jones, Lee Hall, Dan Roberts, N. O. Reynolds, J. A. Brooks, Bill McDonald, James B. Gillett, and John H. Rogers, building a record worthy of the best of them. His years of outstanding service were capped by his capture of killer John Wesley Hardin in 1877. That feat earned him recognition that stood him in good stead the rest of his long life as a businessman, rancher, and civic leader. Chuck Parsons's biography is a long-delayed and much-justified tribute to Armstrong's service to Texas.

The Texas Rangers had their ups and downs in the decades after Armstrong's service. Border war excesses left a stain, and honest Rangers were fired during the political corruption that tainted the state in the 1920s and the early 1930s. Reform governor James Allred cleaned house and rejuvenated the Rangers as part of the Department of Public Safety, giving them back their reputation as a force for the good and allowing them to emerge into the modern law enforcement group we know.

The Rangers would continue to change with the times, taking advantage of new technology as it evolved. In Armstrong's heyday Rangers traveled mostly on horseback or in horse-drawn vehicles, the trips often long and grueling. Later Rangers would take advantage of fast railroad transportation and of the telegraph and telephone, whose messages could far outstrip the fastest getaway horse. They adapted to the automobile age and the constantly improving forensic science they use in crime fighting today.

But no matter the changes, somehow the name "Texas Rangers" always conjures up a popular image of the frontier period in which John B. Armstrong was one of that body's shining lights.

John B. Armstrong and the Travis Rifles

Although no complete roster of the Travis Rifles at the time Armstrong was a member is known to exist, the names of many members have surfaced in newspaper accounts and memoirs. The following is a partial listing (fifty-three of an undetermined number) of those men serving with Armstrong. Brief biographical information is provided if known.

(Spencer J. [?]) Adams (Austin *Daily Democratic Statesman,* October 21, 1874)

Henry B. Barnhart—In 1880 Barnhart was a notary public of the Barnhart & Brumby firm in Austin (*Mercantile,* 51).

C. C. Bell—C. C. Bell worked as a clerk with the William H. Bell grocery firm on Congress Avenue (*Mercantile,* 29).

H. G. Briggs (Austin *Daily Democratic Statesman,* October 21, 1874).

William B. Brush—The 1870 Travis County census shows Brush as a nineteen-year-old "Cook Keeper," the son of Seba (?) Brush, a merchant. Both father and son are shown to be natives of New Jersey (Texas Census, Travis County, enumerated June 1870, by Otto T. Zink, 276). He later worked as a clerk with the firm of S. B. Brush on Congress Avenue (*Mercantile,* 34) and was a member of the Travis Rifles (Barkley, *History,* 233–34).

Charles M. Callaway—Correspondence clerk in the Texas Land Office (*Mercantile,* 36).

Charles B. Cook (Austin *Daily Democratic Statesman,* August 29, 1873).

E. T. Cook (Austin *Daily Democratic Statesman,* August 29, 1873).

Andrew S. Donnan—A. S. Donnan is listed as working as a claims and land agent on Congress Avenue in the 1870s (*Mercantile,* 45).

L. E. Edwards—Edwards's name appears on a roster of the Washington Fire Engine Company of Austin in 1885 (Barkley, *History,* 233–34).

L. H. Fitzhugh (Austin *Daily Democratic Statesman,* October 21, 1874).

William P. Gaines—By the 1880s Gaines is listed as an "attorney and dealer in Texas lands, president Statesman Publishing Co" and as residing at 206 W. Hickory Street, Austin (*Edwards,* 79).

Mr. Gleason—This is possibly Solomon Gleason, in 1870 a family man, born in New York, and living in Austin with his wife, Cecelia, and three children (Texas Census, Travis County, enumerated August 11, 1870, by Otto T. Zink, 218).

Edwin Gray—The son of attorney George H. Gray of Virginia, in 1870 Edwin was clerking in a store and living with his parents and four siblings (Texas Census, Travis County, enumerated June 1870, by Otto T. Zink, 277).

William F. Greene—By 1879 Greene was proprietor of the Club Room Saloon on Congress Avenue in Austin (*C. D. Morrison & Co.'s Directory,* 87; Austin *Daily Democratic Statesman,* June 9, 1878; "Schedule," 44).

I. B. Haber (Weed, "Recollections").

William K. Haralson—Haralson worked as a clerk and boarded in the Durst Boarding House on Brazos Street, where he undoubtedly met and became friends with J. B. Armstrong (*Mercantile,* 56).

Harry Haynes (Austin *Daily Democratic Statesman,* October 21, 1874).

Charles Haynie (Austin *Daily Democratic Statesman,* August 29, 1873).

John P. Holmes—Holmes served as a member of the Frontier Battalion. He also served as a sergeant in the Texas Rangers, joining Company D of the Frontier Battalion under the command of Capt. C. R. Perry. In 1880 he was a twenty-seven-year-old "Commercial Traveler" with a wife, Georgia R., twenty-two, and a one-year-old son, George R. He was from Mississippi (Texas Census, Travis County, enumerated June 10, 1880, by Thomas A. Taylor, 261). Oakwood Cemetery records reveal that he was buried in section 1, lot # 322, on October 5, 1895. His grave is not marked.

(?) Honnett (Austin *Daily Democratic Statesman,* October 21, 1874).

(?) Hopkins (Austin *Daily Democratic Statesman,* October 21, 1874).

John F. House (Austin *Daily Democratic Statesman,* December 27, 1873).

H. F. Kendall (Austin *Daily Democratic Statesman,* August 29, 1873).

H. Levy—Levy worked as a clerk with the firm of Phillipson & Levy, dry goods merchants on Congress Avenue (*Mercantile,* 70).

M. D. Mather—By 1880 Mather was president of Austin's waterworks and living on Congress Avenue. The Austin City Water Company had been incorporated in 1875. Another Travis Rifleman, William B. Brush, was on the board (Barkley, *History,* 245–46). Mather was a native of Vermont, thirty-seven years old, with a wife, R. B., and a one-year-old daughter, Sallie (Texas Census, Travis County, enumerated June 1, 1880, by R. Krause, 172).

Nathan Meyer (Austin *Daily Democratic Statesman,* October 21, 1874).

T. M. Miller—Miller clerked in the firm of Stuart & Mair (*Mercantile,* 78).

Stephen T. Mitchell—Mitchell worked as a clerk with W. R. Thompson, Book Seller, in the Avenue Hotel in the early 1870s (*Mercantile,* 79).

William B. Mitchell—In 1875 W. B. Mitchell was listed as a twenty-five-year-old single male clerking in Fridley's ("Schedule," 38).

John S. Myrick (Austin *Daily Democratic Statesman,* June 8, 1873).

C. W. Orhendorff—Orhendorff worked as a clerk in the firm of William Brueggerhoff, a grocery, on the northeast corner of Congress Avenue (*Mercantile,* 84).

Smith Parr—Parr was twenty-three years old according to the 1875 census, the son of T. W. and R. D. Parr, a lumber merchant ("Schedule," 129). The *Mercantile & General City Directory* shows him as a salesman with the J. W. Wayman mercantile firm on Congress Avenue (86).

J. G. W. Pierson—By 1880 the native-born Texan had relocated to Hamilton, Texas, where he was an attorney and lived with his wife, Eugenia D., and their one-year-old daughter. He was then twenty-nine years old (Texas Census, Hamilton County, enumerated June 18, 1880, by M. S. Brunk, 359).

Albert Samuel Roberts—Roberts served as a member of the Tom Green Rifles, 4th Texas Infantry, Hood's Texas Brigade during the Civil War. In 1870 he earned a commission as a lieutenant in the Texas National Guard. He was also a member of the Travis Rifles (Barkley, *History,* 233–34). He died in Austin on January 11, 1927 (Virginia Roberts Gilman, "Albert Samuel Roberts," in Tyler, *New Handbook,* vol. 3, 607).

R. Robins (Austin *Daily Democratic Statesman,* August 29, 1873).

Alonzo "Lonnie" Robinson—Alonzo Robinson, twenty-one according to the 1875 census, was the son of John H. Robinson ("Schedule," 39). He was born September 19, 1852, and died November 10, 1896. He was a member of the Travis Rifles (Barkley, *History,* 233–34). He is buried in Oakwood Cemetery. He married Laura H. Watson, born August 1, 1853, died September 5, 1915. She is buried beside her husband.

Horace Rowe—Rowe served briefly as a Texas Ranger under Capt. L. H. McNelly. In 1876 he began the publication *The Stylus,* but only five issues were published (Dickey, *Literary Magazines,* 9–10).

Henry Emile Seekatz—According to Seekatz he arrived in Austin on April 1, 1869. A lengthy article, "100 Years Goal of City Pioneer," tells of his shouldering a shotgun and marching along Congress Avenue "to oust Gov. Davis and his regiment of negroes in order for Gov. Coke, who had been elected by the democrats." Seekatz is quoted as saying, "We had blood in our eyes and would have blasted that gang out of the capitol if they hadn't fled before we got there." In 1870 the family resided in New Braunfels, Comal County. Father, Gustav, was a butcher, born in Prussia, as was his wife, Elizabeth. There were five children: Heinrich, Louis, Otto, Emil, and Emma, ranging in age from nine to three years. All the children were born in Texas. In 1880 Henry and Louis were living in Schumansville, Guadalupe County. Henry is listed as being twenty years old and farming. Seekatz did not attain his goal of becoming a centenarian, as he died on February 28, 1940, at age ninety-five (*Austin Statesman,* February 5, 1935). The Travis County Cemetery Record shows that his burial occurred on February 29, 1940, in section 4, lot 840 (Texas Census, Comal County, enumerated June 20, 1870, by Charles Gehren, 124; Guadalupe County, enumerated June 26, 1880, by Francis Gerhard, 282).

(?) Shaw (Weed, "Recollections").

(?) Shelig (Austin *Daily Democratic Statesman,* October 21, 1874).

John M. Swisher Jr.—John M. Swisher Jr. resided with his father and his wife, N. A., twenty-eight, and their two-year-old daughter, Annie E. He is listed as "Rail Road President" ("Schedule," 21). In the early 1870s he clerked in the Department of Education in Austin (*Mercantile,* 104).

Frank H. Tamplet—Tamplet was a bookkeeper with the Sampson & Hendricks merchandising firm in the early 1870s and resided in the City Hotel (*Mercantile,* 105). By 1880 the South Carolina native had relocated to Brenham, Washington County, where he continued as a bookkeeper and lived with his wife, Sallie, his mother-in-law, and three brothers (Texas Census, Washington County, enumerated June 5, 1880, by J. James, 120).

Arthur Terrill—The 1875 census shows the eighteen-year-old Terrill living in the household of A. H. Cook Jr. ("Schedule," 58).

Sam Wade—Samuel Wade was seventeen years old, clerking in a law office, and boarding in the household of M. Lu ("Schedule," 22). He was certainly the youngest of the Travis Rifles.

David Walker—In 1875 Walker was a single twenty-eight-year-old and living in the City Hotel ("Schedule," 18). He earlier had worked as a night clerk in the Raymond House on the corner of Congress Avenue and Pine Street (*Mercantile,* 110). The March 14, 1877, Austin *Daily Democratic Statesman* notes his death; his remains were escorted to Oakwood Cemetery by the Austin Greys and the Travis Rifles.

Charles H. Webb—Webb worked with A. S. Donnan in the latter's Claim & Land Agents firm (*Mercantile*, 111).

V. O. Weed—Valentine Osborn Weed began in the livery business and then entered the funeral business. He died August 8, 1935, in Austin (*Austin Statesman*, August 8, 9, 1935).

(?) Whiting—Possibly R. B. Whiting, who in the 1875 census is listed as a clerk in a law office and living with his parents, two younger siblings, and four unnamed children ("Schedule," 60).

(?) Wildy (Austin *Daily Democratic Statesman*, October 21, 1874).

Dr. Edward Wise (Austin *Daily Democratic Statesman*, June 14, 1873).

Charles C. Worthington—Worthington was twenty-one years old in 1875, the son of Dr. C. D. and L. H. Worthington ("Schedule," 32). The Austin *Mercantile & General City Directory* of 1872 shows him as a clerk in the Savings Bank of Austin (114).

Edward B. Wright—The Rev. Wright was a successful minister. The 1880 Travis County federal census shows him as a forty-two-year-old Presbyterian minister, born in Ohio. His wife, Evelyn, thirty-one, and seven-month-old child were born in Texas (Texas Census, Travis County, enumerated June 3–4, 1880, by Fritz Tegener, 206). He became pastor of the First Presbyterian Church of Austin in 1872 and served in that position for over forty years (Barkley, *History*, 291).

APPENDIX B
Serving with McNelly

The following is the first roster of the Washington County Volunteer Militia to carry Armstrong's name. He began as a private and was promoted to fifth sergeant on May 20, 1875. This roster was prepared at Santa Maria, Cameron County, August 31, 1875, and shows the names of those mustered in on April 1, 1875, unless otherwise noted.

Capt. L. H. McNelly
Lt. T. C. Robinson
Lt. James W. Guynn (June 22)
Sgt. George A. Hall (January 20)
Sgt. Roe P. Orrell

Sgt. Charles M. Middleton
Sgt. Lawrence Baker Wright
Sgt. Linton Lafayette Wright
Cpl. J. Brown (April 6)
Pvt./Clerk J. R. Wofford (April 6)

PRIVATES

S. J. Adams
G. H. Allen (June 22)
John B. Armstrong (May 20)
G. Blackford (June 14)
E. Brown (June 22)
F. Bolton (June 18)
W. E. Bridge (June 22)
William C. Callicott
W. H. Cox (June 22)
R. Dance (June 22)
C. W. Dupriest
George P. Durham
M. Fleming
W. D. Glass
J. M. Harrison
C. W. Howery
M. M. Halyard (June 22)
A. Hagan (June 22)
J. Hickey (June 22)
G. Jarrells (June 22)
J. Jarrells (June 22)
M. Jones
R. Jones (June 22)
A. S. Lowery (June 22)

Horace Maben/Mabin
G. Mathews (June 22)
E. McGinnis (June 22)
Thomas McGovern (May 1)
A. S. McKay (June 22)
A. J. McNeil (June 22)
C. Moore (May 1)
G. Moore (May 1)
Thomas Melvin (July 22)
Pedro Mayniel (July 1)
J. S. Napier (July 22)
Shadrach M. Nichols
E. A. Northington
R. A. Petty (May 17)
R. H. Pitts
J. Racy (June 22)
H. G. Rector
José Ramírez (July 1)
William L. Rudd
F. Siebert (July 6)
G. M. Scott (April 6)
Jesús Sandoval (May 1)
D. R. Smith
E. F. Shaw (June 22)

Felipe de los Santos (July 1)   M. F. Welch
William R. Templeton (April 6)   M. H. Williams
J. N. Took (June 22)   B. T. Wilks (June 22)
Reyies [Reyes?] Vela

DIED

Pvt. L. B. Smith—"Killed in action with Mexicans," June 12, 1875

DESERTED

Pvt. J. Fartheree (May 20)
Pvt. J. Williams (May 20)

DISCHARGED — OFFICERS

Cpl. J. E. Smith (served April 6 –June 30)
Cpl. W. M. Taylor (served April 1–June 30)

DISCHARGED — PRIVATES

W. F. Alderson (served April–May 14)
J. Bartholow (served May 1–June 14)
A. Brown (served April 6 –June 30)
V. B. Byers (served June 22–July 14; "Dishonorably discharged; Could not comply with Company rules")
J. B. Carter (served June 22–July 14; "Dishonorably discharged; Could not comply with Company rules")
L. R. Carter (served June 22–July 14; "Dishonorably discharged; Could not comply with Company rules")
M. Dupriest (served April 1–May 14)
E. D. Howland (served June 22–July 14; "Dishonorably discharged; Could not comply with Company rules")
J. M. Norton (served April 1–July 31)
J. N. Patterson (served April 10 –May 19)
C. Perry (served June 22–July 14; "Dishonorably discharged; Could not comply with Company rules")
Oscar F. Pridgen (served April 6 –June 1)

Serving under Hall

Jesse L. Hall took command of McNelly's troop on January 1, 1877, but his first muster and payroll is dated January 25. With McNelly out, the Washington County Volunteer Militia Company now became known simply as the Special State Troops. The following "Muster and Pay Roll of Lieut. J. L. Hall Company Specl State Troops in the Service of the State of Texas, from Jany 25th 1877 to March 31, 1877 mustered into service Jany 25th 1877" is the first roster under Hall's command which carries Armstrong's name. This muster roll was prepared at Clinton, DeWitt County, March 31, 1877.

2nd Lt. J. L. Hall (pay period January 25–March 31; received $125 per month)

1st Sgt. J. B. Armstrong (pay period January 25–March 31; received $50 per month. As of December 1, 1877, Armstrong held the rank of first lieutenant, receiving $133 per month)

2nd Sgt. O. S. Watson (pay period January 25–March 31; received $50 per month. The last payroll shows him serving from August 31 to October 31, 1877, when he was discharged)

Cpl. A. L. Parrott (pay period January 25–March 31; received $40 per month)

Cpl. N. R. J. Stegall (pay period January 25–March 31; received $40 per month)

PRIVATES
(received $40 per month) mustered in January 25, 1877, unless otherwise noted

G. H. Allen (discharged November 30, 1879)
C. T. Allen
C. W. Covington (served March 1–31, 1877)
W. T. Davis
T. W. Deggs
George P. Durham
O. R. Erwin (served March 1–31, 1877)
W. H. Griffin
S. N. Hardy (discharged November 30, 1877)
C. W. Howery
D. W. Holden (served February 17–March 31, 1877)
E. R. Jenson (discharged October 15, 1878)
Napoleon A. Jennings (discharged April 30, 1877)
Horace Maben/Mabin
A. S. McKay
William W. McKinney (discharged June 1, 1878)
Charles Brown McKinney

F. H. Miller (served March 24; discharged May 1, 1877)
T. B. Menefee (served March 24; discharged June 1, 1878)
Samuel A. McMurray (served March 24–31, 1877)
J. F. Pendleton (served March 24–31, 1877)
H. G. Rector (discharged October 15, 1878)
William L. Rudd
George W. Talley (discharged August 31, 1878)
Linton L. Wright (served February 15–March 31, 1877)
H. C. Wilson (served March 24–31, 1877)
Alfred Walker, teamster (received $30 per month)

Texas Volunteer Guard

Three rosters of the Texas Volunteer Guard that carry Armstrong's name are known. The earliest is dated April 3, 1889, at Brownsville and is entitled "Muster Roll of Company G First Regiment Texas Volunteer Guard." This shows Armstrong mustering in as a private on April 3, 1889, his occupation given as "Stockman." He is shown to be thirty-eight years old, with blue eyes, light-colored hair, and a fair complexion. The unit was commanded by Capt. J. F. Cummings, who mustered in on February 14, 1889. His occupation is given as "Teacher."

OFFICERS

Capt. J. F. Cummings
1st Lt. M. Leahy
2nd Lt. A. B. Crane
1st Sgt. H. C. Chenoweth
Sgt. B. A. Turegano
Sgt. M. Hanson Jr.
Sgt. J. L. Putegnat

Sgt. William Kelly Jr.
1st Cpl. J. G. Stucke
Cpl. A. W. Cowen
Cpl. O. C. Sander
Cpl. C. H. Thorn
Cpl. F. N. Cowen

PRIVATES

J. B. Armstrong
A. B. Barton
J. H. Bloomberg
A. A. Browne
F. Champion
G. Champion
George Conner
A. B. Cowen
L. W. R. Cowen
R. M. Dalzell
S. L. Doorman
B. Escatiola
J. Marks
J. F. Mason

C. A. Michel
Frank Moore
George More
W. A. Neale
G. M. Putegnat
H. S. Putegnat
G. L. Rendall
W. J. Russell
J. P. Scanlon
W. J. Scanlon
A. Thornham
J. Thornham
C. F. Tilghman
A. S. Wolff

DISCHARGED

1st Lt. L. E. Chanay
Pvt. C. B. Combe Jr.
Pvt. A. Celaya
Pvt. R. C. Macy

Pvt. C. Marlin
Pvt. G. W. Miller
Pvt. Frank Tenille
Pvt. A. T. Woodhouse

Armstrong Family Tree

Gen. Martin Armstrong (c. 1737–1808) m. Mary Elizabeth Tate (c. 1747–1836)

Dr. John Barkley Armstrong Sr. (1783–1836) m. Mary Turner (1790–1854)

Dr. John Barkley Armstrong Jr. (1819–1875) m. Maria Susannah Ready (1813–1885)

John Barkley Armstrong [III] (1850–1913) m. Mary Helena "Mollie" Durst (1855–1897)

Maria Josephine
(1879–1972)
m. Andrew Stewart

Jamie Durst
(1881–1963)
m. John M. Bennett

John Barclay
(1884–1905)

Charles Mitchell
(1886–1941)
m. Lucie Tobin Carr

Julia Katherine
(1889–1991)
m. Zeb Mayhew;
Allard Kaufmann

Elliott Ropes
(1890–1898)

Thomas Reeves
(1892–1986)
m. Henrietta K. Larkin

John Barclay
Charles Mitchell, Jr.
Tobin (1923–2005) m. Anne Legendre

John Barclay
Katharine Armant
Sara Storey
Tobin, Jr.
James Legendre

Source: Smith, *Armstrong Chronicle*; partial family tree in Tobin Armstrong Collection

Notes

PREFACE

1. Haley, *The XIT Ranch,* 123–24.

2. Preece, *Lone Star Man,* 234.

3. Trachtman, *The Gunfighters,* 201.

4. Aten, "Six and One Half Years," 100.

5. From an unidentified newspaper clipping, "Delegation from Dalhart Big Part of Gathering." This article describes the first XIT reunion held in Fort Worth. Here Armstrong's slayer is identified as Gene *Elliston,* not Ellison.

CHAPTER 1

1. Smith, *Armstrong Chronicle,* 333. Dr. John Barkley Armstrong was born January 20, 1819, in Giles County, Tennessee, and was married to Maria Susannah Ready on May 23, 1842. Maria Ready's father founded Readyville, Tennessee. The Armstrongs' children were Thomas Temple, born February 7, 1843; Mary "Mollie" Ready, born November 18, 1844; William Francis, born September 12, 1846; Laura Maria, born September 16, 1847; John Barkley, born January 1, 1850; Lavanda "Van" Martin, born September 10, 1852; and Betavia "Beta" Jane, born November 13, 1854. All lived to adulthood except William Francis, who died October 3, 1846. Laura Maria died July 28, 1870. All the other children survived well into the twentieth century. Dr. Armstrong died December 18, 1875, in McMinnville. I determined Readyville as Ranger Armstrong's birthplace from his "Descriptive List," which had to be carried for identification by all Rangers.

2. U.S. Bureau of the Census, Census for Tennessee (hereafter Tennessee Census), Cannon County (free), enumerated October 28, 1850, 430. Dr. Armstrong owned only three slaves at this time (Tennessee Census), Cannon County (slave), enumerated October 9, 1850, 117.

3. Numerous references to Armstrong, including his official death certificate, spell his middle name "Barclay." I use "Barkley" as that is the spelling found in the family bible, Tobin Armstrong Collection.

4. Tennessee Census, Warren County, enumerated August 30, 1860, 516. In December 1856 or January 1857, Dr. Armstrong moved his family and slaves to McMinnville (Smith, *Armstrong Chronicle,* 118). By 1860 Dr. Armstrong owned seven slaves (slave inhabitants in McMinnville, Warren County, Tennessee, Tennessee Census, enumerated August 24, 1860, 422).

5. Undated interview with Tobin Armstrong at the Armstrong ranch; Smith, *Armstrong Chronicle,* 119–20.

6. Smith, *Armstrong Chronicle,* 120–21.

7. Ibid., 121. Smith states that Armstrong "presented himself to Austin" on January 1, 1873, with but $2.50 in his pocket.

8. In the early 1870s Mather was with the Reed & Mather furniture business, located on the corner of Congress Avenue and Ash Street (now 9th Street).

9. Albert S. Roberts became a successful wholesale and retail grocer in Austin after serving as a member of the Tom Green Rifles, 4th Texas Infantry, Hood's Texas Brigade during the Civil War. By 1870 he had been commissioned a lieutenant in the Texas National Guard and in that capacity was in command of the Austin Company of the Travis Rifles. His company took over the capitol during the conflict between the defeated governor E. J. Davis and newly elected governor Richard Coke, January 15–16, 1873. He retired on July 1, 1895, with the rank of major general. In civil life he served as a postal inspector. He died in Austin on January 11, 1927 (Gilman, "Albert Samuel Roberts," in Tyler, *New Handbook,* vol. 3, 607; *Austin American,* January 12, 1927). Andrew S. Donnan operated A. S. Donnan & Co. Claim and Land Agents on Congress Avenue in the early 1870s (*Mercantile,* 45). Charles H. Webb was a member of the A. S. Donnan & Co. Claim and Land Agents firm (*Mercantile,* 111). In 1870 William B. Brush was a nineteen-year-old "Cook Keeper" working for his father, S. B. Brush, a merchant in general merchandise in Austin. Both were natives of New Jersey (Texas Census, Travis County, June 1870, 276; *Mercantile,* 34). Swisher's father was a clerk for the Auditorial Board, Treasury Building. His son, a Travis Rifleman, clerked in the Department of Education in Austin in the early 1870s (*Mercantile,* 104). T. M. Miller clerked in the firm of Stuart & Mair (*Mercantile,* 78; *Daily Democratic Statesman,* June 8 and 14, 1873; Brown, "Annals of Travis County, 1873," 15).

10. *Daily Democratic Statesman,* August 20, 1873.

11. Henry B. Barnhart's name is frequently seen spelled as *Barnhardt.* I use *Barnhart* as that is how it is spelled in the Austin city directories as well as on his headstone in Oakwood Cemetery in Austin. In 1877–78 he worked as an attorney and notary public, with an office on the northwest corner of Congress Avenue and Hickory Street (present-day 8th Street) (*Mooney & Morrison,* 69). Edwin Gray was the son of attorney George H. Gray from Virginia. In 1870 Edwin was a clerk in a store and living with his parents and four younger siblings (Texas Census, Travis County, enumerated June 1870, 277). Stephen T. Mitchell was a clerk with W. R. Thompson, Book Seller, in the Avenue Hotel in the early 1870s (*Mercantile,* 79). Horace Rowe served briefly as a Texas Ranger under Capt. L. H. McNelly, as did Armstrong. In 1876 Rowe began the periodical *The Stylus: A Monthly Magazine Devoted to Literature, Science and Art,* but only five issues were published (Dickey, *Literary Magazines,* 9–10). Valentine Osborn Weed, born about 1850, for many years was in the livery business in Austin before entering the funeral home business. He died August 8, 1935, at his residence in Austin (*Austin Statesman,* August 8 and 9, 1935). Weed's descendants still operate the funeral establishment his father, Thurlow Weed, began, now called Weed-Corley-Fish. Charles C. Worthington was the son of Dr. Charles and Louise H. Worthington of Austin. Sixteen years old in 1870, he was among the youngest of the Travis Rifles. He later worked as a clerk in the Savings Bank of Austin (*Mercantile,* 114).

12. *Daily Democratic Statesman,* December 27, 1873.

13. *Mooney & Morrison,* II. In 1860 Mary Josephine Durst, now a widow, her husband having succumbed to an illness on April 24, 1858, was living with her relatives, the William Woods Atwood family. Atwood was a successful stock raiser with $6,000 in real estate and a $3,000 personal estate. The widow Durst had two children, Mortimer T. and Mary Helena, and is shown in the census to have $2,320 real estate and $15,100 worth of personal estate (Texas Census, Travis County, enumerated July 24, 1860, 256; Smith, *Armstrong Chronicle,* 59).

14. George B. Zimpelman was first elected Travis County sheriff on June 25, 1866, and served until November 1, 1867, when he was removed by Gen. J. J. Reynolds's Special Order No. 195. He was again elected on December 3, 1869, reelected on December 2, 1873, and served

until February 15, 1876 (Tise, *Texas County Sheriffs,* 494). His last name is frequently seen as "Zimpleman," but I have used the spelling which appears on his large headstone in Oakwood Cemetery, Austin. Zimpelman lived from July 24, 1832, until January 1, 1909. He is buried beside his wife, Sarah Catherine, who lived from March 7, 1837, to November 12, 1885.

15. *Daily Democratic Statesman,* January 16, 1874. Also see "Travis Guards and Rifles," in Tyler, *New Handbook,* vol. 6, 555. The typed copy of Frank Brown's *Annals* in the Austin History Center spells the captain's name as *Mather* whereas the *New Handbook* spells it *Mathew.* Mather is correct, as later *Daily Democratic Statesman* articles (June 6, October 21, 1874) use that spelling, as does F. L. Britton's *Report of the Adjutant General of the State of Texas for the Year 1873.* In "Statement of Ordnance and Ordnance Stores, Pertaining to the State of Texas, December 31, 1873" (part of Britton's 1873 report), M. D. Mather is listed as captain of the Travis Rifles stationed in Austin.

16. Weed, "Recollections." This is a six-page manuscript prepared by Weed for W. S. Red on November 20, 1930. At the time, Weed identified I. B. Haber, W. K. Haralson, Harry Haynes, and himself as the only living members of the Travis Rifles of 1873. His participation in the recovery of the capitol building for newly elected governor Richard Coke is mentioned in his obituary. Weed was also a Tennessee native.

17. *Daily Democratic Statesman,* January 17, 1874.

18. Ibid., January 27, 31, 1874. Quotation from January 31 issue.

19. Ibid., February 14, 1874.

20. Ibid., June 6, 1874.

21. The L. H. Fitzhugh family operated a boardinghouse on Bois d'Arc Street (now 7th Street); see "Schedule," 38.

22. *Daily Democratic Statesman,* June 23, July 1, 1874.

23. Ibid., October 21, 1874. Although no complete roster of the Travis Rifles is known to exist, this newspaper report does give the names of thirty-two participants in the contest. See appendix A.

24. In 1875 "Salom" Smith was residing in the household of Sarah R. and Robert A. Smith. She was single and thirty-one years old. Robert A. Smith was thirty-four and perhaps her older brother. He was an attorney with an office on Congress Avenue ("Schedule," 47). Her grave in Oakwood Cemetery in Austin is marked with a simple stone reading "Salome Smith / 1844–1876." Annie DeCordova was born about 1856, the daughter of J. D. and Phineas DeCordova. Her father was a successful land agent ("Schedule," 86). *Daily Democratic Statesman,* December 6, 1874.

25. *Daily Democratic Statesman,* December 6, 1874. Miss Haralson had earlier been chosen as the Queen of Beauty at the equestrian exhibit of the Travis Rifles. The Committee of Arrangements included C. C. Bell, H. G. Briggs, L. E. Edwards, and A. S. Roberts.

26. For histories of the Texas Rangers focusing on the nineteenth century, see Robinson, *Men;* Utley, *Lone Star Justice;* and Wilkins, *Law.* For a concise study of the change in the role of the Rangers from the pre–Civil War period and the late nineteenth and the early twentieth centuries, see Weiss, "Texas Rangers."

CHAPTER 2

1. McNelly's muster and payroll, prepared at the Magotee de Don Juan, August 31, 1875 (Texas State Library and Archives, Austin), shows that Armstrong had served as a private for

three months (May 20–August 20) at $13.00 per month for a total of $39.00, and as fifth ser-geant for eleven days, August 20–31, at $17.00 per month, for a total of $6.23. In addition, he received a clothing allowance of $4.00 per month for $13.33, and fifty cents per day horse al-lowance ($50.50), for a total of $109.06. These documents are now in the Texas State Archives. In 1875 a private in McNelly's company received $13 per month, a corporal, $15 per month, and a sergeant, $17 per month.

2. Cynthia E. Orozco, "Nuecestown Raid of 1875," in Tyler, *New Handbook,* vol. 4, 1059. Today all that remains of Nuecestown is a schoolhouse built in 1892 and the Nuecestown ceme-tery, all within the city limits of Corpus Christi.

3. McClane's telegram, received April 18 at Steele's office, is in the Adjutant General Papers. Steele's response is dated April 19 and is found in the Adjutant General's letter press book, Ad-jutant General Papers. McNelly reported that he left on April 10; Steele reported that he left on the ninth. The sheriff's name is occasionally spelled as *McClure,* although he signed his name clearly as *McClane.*

4. Jennings, *Texas Ranger,* 63. Although Jennings telescopes some events and placed him-self in other actions in which he could not have participated, he did serve under McNelly from May 26, 1876, until his honorable discharge on February 1, 1877.

5. Durham, "5100 Outlaws," part 1, 111. This was later published in book form under the title *Taming the Nueces Strip* with slight variations. Interestingly, at one time Clyde Wantland, to whom Durham told his story, was collecting information to write a biography of Armstrong. The July 1941 issue of the *Southwestern Historical Quarterly* carried this item in the "Texas Collection" section: those "who have information about John B. Armstrong may communicate with Mr. Wantland at [his address]." At the time, Wantland was editor of the *Alamo Heights News.* It is not known if he ever got past the idea stage of writing the biography

6. John B. Armstrong, "Descriptive List." The note on the form states, "This Descriptive List, for identification, will be kept in possession of the Ranger to whom it refers, and will be exhibited as a warrant of his authority as such, when called upon, and must be surrendered to his Company Commander when discharged." Original document in Armstrong's Service Rec-ord file, Texas State Archives.

7. This notice appeared in the *Galveston Daily News* of May 4, 1875, as well as in the *San Antonio Daily Herald* of May 6; it was reprinted in other newspapers as well.

8. For a thorough discussion of McNelly and his handling of these Minute Men compa-nies, see Parsons, *Captain L. H. McNelly,* 184–86.

9. Durham, "5100 Outlaws," part 2, 113. Curiously, the speech is not mentioned in Dur-ham's *Taming the Nueces Strip.* Durham later became a *caporal* (foreman) on the King Ranch. He died in Willacy County on May 28, 1940.

10. *San Antonio Daily Herald,* May 17, 1875.

11. Durham, "5100 Outlaws," part 2, 113. Little is known of Lino Saldaña. He was born in Texas about 1817. By 1880 he was farming in Cameron County, living with his wife, Jesusa, age fifty-four, and a fifteen-year-old son, Manuel. His wife and son were both born in Mexico (Texas Census, Cameron County, enumerated June 16, 1880, by Joseph P. O'Shaughnessy, 453). O'Shaughnessy was at this time city marshal of Brownsville.

12. McNelly's report was originally published in the *Colorado Citizen,* the Columbus news-paper in Colorado County, on July 1, 1875. It was probably sent to the editor by Lt. J. W.

Guynn, who was with McNelly and from Colorado County. For a thorough discussion of Mc-Nelly's Palo Alto Prairie victory, see Parsons, *Captain L. H. McNelly,* 195–206.

13. Unfortunately, McNelly provides no list of the men who actually participated in this engagement. From the several accounts left by Rangers who were there a partial list can be constructed. Besides McNelly there were Lt. T. C. Robinson; Jesús Sandoval; William C. Callicott; L. B. Smith; his father, D. R. Smith; Spencer J. Adams; Linton L. Wright; Lawrence B. Wright; Thomas Sullivan; George Durham; William Rudd; Roe P. Orrell; Horace Maben (or Mabin); and, of course John B. Armstrong.

14. Testimony of McNelly before the committee dealing with Texas border troubles, given January 29, 1876 (U.S. Congress, *Texas Frontier Troubles,* 17 [hereafter cited as *Texas Frontier Troubles*]).

CHAPTER 3

1. *Texas Frontier Troubles,* 8–9.

2. Ibid., 14–15.

3. Robinson's letter, signed with the pseudonym "Pidge," was printed in the *Daily State Gazette* on January 19, 1876. His complete writings have been printed in full in Parsons, *"Pidge."* The "invasion letter" is found on 87–91.

4. McNelly to Adj. Gen. William Steele, from Brownsville, November 12, 1875, Adjutant General Papers.

5. *Daily State Gazette,* January 19, 1876. Rancho Las Cuevas today is the city of Gustavo Díaz Ordaz, Tamaulipas, Mexico.

6. Ibid.

7. George A. Hall was no relation to Jesse L. Hall, who would soon become commander of the McNelly troop; George was related to McNelly through marriage. McNelly married Carey Cheek Matson, the daughter of Sarah Hall and John Cheek. A. C. Hall, a brother of Sarah, was the father of George A. Hall. On January 20, 1875, George A. Hall was mustered into McNelly's militia company and served until February 1, 1877. In November 1877 G. A. Hall was elected captain of a Fayette County militia organization. He died at his home in La Grange, Fayette County, on Sunday morning, October 23, 1921. His obituary points out that "during his early life [Hall] served as a Texas Ranger, ever fearless" (Hall's Ranger service record; muster and payroll of Fayette County Militia, Texas State Archives; and *La Grange Journal,* October 27, 1921).

8. William Crump Callicott also left an account of the invasion, writing it for Walter Prescott Webb to be utilized in *The Texas Rangers: A Century of Frontier Defense.* Callicott was almost blind when he wrote his account not only of the invasion but of other experiences while a Texas Ranger. Because he hadn't read much of events written by others, he is very reliable, as he was working solely from memory of his own experiences; his memory was not "tainted" by the writings of others. Callicott served under McNelly from April 1, 1875, to February 15, 1876. He died in Houston on June 10, 1926, and is buried in the Magnolia Cemetery there. The Callicott papers, including photographs, letters, and his large hand-drawn map of the Palo Alto Prairie battlefield, are currently archived at the Center for American History, Walter Prescott Webb Collection, Austin. Hereafter cited as Callicott correspondence.

9. *Daily State Gazette,* January 19, 1876.

10. Ibid.

11. For a complete description of McNelly's invasion of Mexico and subsequent events, see Parsons, *Captain L. H. McNelly,* 159–257.

12. McNelly to Steele, November 21, 1875, from Ringgold Barracks, Adjutant General Papers.

13. Callicott correspondence, undated manuscript.

14. According to George Durham, the men were McNelly; Robinson; Sergeants Hall and Armstrong; Cpl. William Rudd; Pvts. William Callicott, Horace Maben (Mabin), Tom Mc-Govern, R. H. Pitts, and Jim Wofford; and himself (Durham, *Strip,* 122–23). Durham probably served continuously from his enlistment on April 1, 1875, to June 1, 1877, although the existing service records show he served under McNelly from April 1 to August 31, 1877, and from January 25, 1877, to June 1, 1877 under Hall.

15. McNelly to Steele, November 21, 1875, from Rio Grande City, Starr County, Adjutant General Papers.

16. Callicott correspondence, undated manuscript, 114–16.

CHAPTER 4

1. T. C. Robinson to Adjutant General Steele, December 30, 1875. This appears in slightly different form in Steele, *Report* (9): "On 28th, a scouting party came across a slaughter-house for stolen beeves, about forty miles north of Las Kuscias [Rucias]. Ranchero in charge arrested. After an ineffectual attempt to bribe the sergeant, he tried to escape and was killed in the attempt." By now Lieutenant Robinson was performing most of McNelly's "paperwork" duties.

2. During Armstrong's day this community was known as Edinburgh and was the county seat of Hidalgo County. In the 1850s John Young, a Scotsman, settled there and renamed the community after the city in Scotland. A post office was established in 1876, but in 1885 the name was changed to Hidalgo. In 1908 the county seat was moved to Chapin, which was then renamed Edinburg, without the "h" (Alicia A. Garza, "Hidalgo, Texas," in Tyler, *New Handbook,* vol. 3, 589).

3. Steele, *Report, 1876,* 9.

4. The only full-length biography of Fisher remains Fisher and Dykes, *King Fisher: His Life and Times.* E. Lee Fisher has produced a well-written novel based soundly on Fisher's life entitled *Pendencia Creek: The Life and Times of a Texas Gunfighter.*

5. McNelly to Steele, June 4, 1876, Adjutant General Papers. Curiously, William R. Templeton had served under Captain McNelly as a private from April 6 through October 31, 1875. A letter in his service file from Adj. Gen. W. W. Sterling dated August 19, 1932, gives his dates of service as April 6 to October 1 (Adjutant General Papers). The monthly return of October 31, 1876, shows he was discharged on that date "at his own request" but owed the state for one Sharps carbine and a Colt pistol. Templeton died in Blanco County on March 30, 1911 (Adjutant General Papers).

6. *San Antonio Daily Express,* June 6, 1876.

7. Steele, *Report,* 9; Jennings, *Texas Ranger,* 111–15. Jennings was born in Philadelphia on January 11, 1856, and spent some time in Texas as a young man. His service record shows that he served under Captain McNelly from May 26, 1876, to February 1, 1877, and also under Lt. Jesse L. Hall from January 25 to April 30, 1877, when he was honorably discharged (Adjutant General Papers). Jennings died on December 15, 1919, in New York City.

8. Steele, *Report,* 9.

9. Affidavit of L. B. Wright, Adjutant General Papers.

10. Steele, "Results," 2. Edward R. Garner was the son of Matilda and J. I. Garner and was born in Mississippi about 1851. He qualified for sheriff on May 6, 1874, when he was about twenty-three, and served until February 15, 1876 (Tise, *Texas County Sheriffs;* Texas Census, San Patricio County, enumerated August 21–22, 1870, by R. A. Upton, 99–100).

11. Captain McNelly monthly return, September 30, 1876, prepared at Oakville, Live Oak County (Adjutant General Papers).

12. The spelling is frequently seen as *Espenosa* but the official spelling is *Espantosa* (which means "frightening"). A description of the lake written at the time that concerns us does not hint at the ghostly quality of the water: "Lake Espantosa, with its arms, must be seven or eight miles in length. Though narrow, it is immense in depth, and contains almost every variety of fish. Its banks are lined with live oak, which are heavily laden with the creamy mast. Mr. John McElroy, the gentleman proprietor of the rancho near by, has a fine lot of swine to eat them" (*Galveston Daily News,* October 24, 1878, reporting news from Dimmit County).

13. The August 15, 1876, bank robbery, carried out by nine disguised men, created considerable publicity, but as far as is known no one ever stood trial for the robbery; see the *Victoria Advocate,* August 17, 1876. In November Frank Callison, having been arrested by Frio County citizens and turned over to Bexar County sheriff William B. Knox, confessed to Lt. J. L. Hall that he had taken part in the bank robbery and named Tom Callison, William Cavin, Lark Ferguson, James T. Trimble, Alf Day, John Grun, John Cabler, and Thomas Jasper as accomplices. He further named King Fisher, William Taylor, "Dock" Cornett, and William Brooking as men who aided and abetted (Hall to Steele, November 4, 1876, Adjutant General Papers). Apparently, the confession led nowhere.

14. Jennings, *Texas Ranger* (123), provides the names of the Rangers with Armstrong.

15. M. H. Williams's service record shows he was mustered into McNelly's troop on April 1 and served until August 31, 1875, and from December 1, 1876, to February 1, 1877. He served under Hall in the Special State Troops from August 1 to November 30, 1879 (Adjutant General Papers).

16. Devine's account, written on September 4, 1877, is found in Hubbard correspondence, record group 301, Texas State Archives. Devine joined McNelly on August 1, 1876, and served until December 14, 1876, a period of only four months. After his brief career as a Ranger he practiced law in Bexar County. By 1880 he was an attorney-at-law and had a wife, Mary, and a daughter, Helen Rose (Texas Census, Bexar County, enumerated June 29, 1880, by Max Neuendorff, 107). Devine later became judge of the City Corporation Court and justice of the peace. After a lengthy illness he died on May 9, 1929 (*San Antonio Daily Express,* May 10, 1929).

17. Account of Thomas N. Devine, September 4, 1877, Adjutant General Papers.

18. Armstrong's account is in the form of a telegram to McNelly, dated October 1. It was printed in full on page one of the October 3, 1876, issue of the *Daily Democratic Statesman* of Austin, the *Galveston Daily News,* and the *San Antonio Daily Express.* The *Daily Democratic Statesman* headlined its article "McNelly and Desperadoes"; the *Galveston Daily News,* "Death to Desperadoes"; and the *San Antonio Daily Express,* "McNelly Strikes Again." Jennings, "with Armstrong," wrote the report of the fight. Steele's report ("Results," 2) states the number of stolen horses recovered as fifty and cattle as thirty-two head.

19. Jennings writes that it was Cpl. William Rudd who killed the "bad Mexican" named Pancho Ruiz, wanted for murder in Corpus Christi (*Texas Ranger,* 127).

20. Armstrong to McNelly, October 1, 1876; see note 18 above.

21. *San Antonio Daily Express,* October 3, 1876.

22. McNelly's monthly return, October 31, 1876, prepared at San Antonio (Adjutant General Papers). At this time, McNelly's illness was causing him to spend more and more time in San Antonio working with doctors. Armstrong was now, in effect, in charge of the Special State Troops.

23. *San Antonio Daily Express,* October 20, 1876, reprinting an undated item from the *Pleasanton Stock Journal.*

24. *San Antonio Daily Express,* October 28, 1876. The issue of October 31 dutifully notes that Armstrong and Horace Rowe, "McNelly's Rangers," were staying at the Menger Hotel; the issue of November 14 notes that J. B. Armstrong and T. W. Deggs of "McNelly's Co." were staying at the Central Hotel.

25. McNelly's monthly return, October 31, 1876, prepared at San Antonio (Adjutant General Papers). The jail was called the "Bat Cave" because of the myriad of bats that roosted in the ceiling and eaves of the jail, the courthouse, and city hall.

26. Dr. E. Melon to Steele, November 15, 1876 (Adjutant General Papers).

27. McNelly to Steele, November 13, 1876, from San Antonio (Adjutant General Papers); McNelly's monthly return, November 30, 1876 (Adjutant General Papers). As McNelly's health worsened he spent more and more time in San Antonio closer to his doctors. He trusted his lieutenants and sergeants to perform quality work during his absence.

28. Reward proclamation printed in the *Daily Democratic Statesman,* January 22, 1874. The reward proclamation by Gov. Richard Coke is dated January 20, 1874.

29. *San Antonio Daily Express,* December 7, 1876. "Killed for Resisting Arrest" ran the headline. Graytown, originally in Bexar County, was placed in Wilson County by a boundary change in 1869. The community diminished in size until, in the 1990s, it "was a dispersed community with a population of sixty-four" (Minnie B. Cameron, "Graytown, Texas," in Tyler, *New Handbook,* vol. 3, 300).

30. Jennings, *Texas Ranger,* 138. The *San Antonio Daily Express* (December 7, 1876) reported that Mayfield was mounting a horse when the Rangers ordered him to surrender.

31. Armstrong to Steele, December 9, 1876, written from San Antonio (Adjutant General Papers). Robert Montgomery was an elderly man with a family when he was killed. The 1870 census shows him to be a seventy-four-year-old farmer, born in North Carolina (Texas Census, Parker County, enumerated August 22, 1870, by L. M. Parker, 386).

32. *San Antonio Daily Express,* December 7, 1876.

33. *Galveston Daily News,* December 12, 1876, reprinting an article from the *San Antonio Daily Herald,* December 8, 1876.

34. *San Antonio Daily Express,* December 13, 1876.

35. Jennings, *Texas Ranger,* 138.

36. Lt. L. B. Wright to Steele, December 10, 1876, written from San Antonio (Adjutant General Papers). Jennings wrote that Lieutenant Wright took Armstrong and Deggs and ten other men to investigate the Mayfield killing.

37. Armstrong to Steele, December 9, 1876, written from San Antonio (Adjutant General Papers).

38. Steele to Armstrong, December 15, 1876 (Adjutant General Papers).

39. Jennings, *Texas Ranger,* 138.

40. Copy of the Lindsey letter provided by Mrs. Mary M. McClurg, the great-great-granddaughter of the slain Robert Montgomery and the great-granddaughter of J. N. Montgomery, May 7, 1977.

41. *Galveston Daily News,* December 22, 1876, citing a special telegram from San Antonio; McNelly's monthly return, prepared at Clinton, DeWitt County, December 31, 1876.

42. The family name has been spelled in a variety of ways. The spelling I use here is from their headstones in the Upper Yorktown, DeWitt County, Cemetery.

43. For a thorough discussion of the arrest of the seven men, see Parsons, *Captain L. H. Mc-Nelly,* 281–89.

CHAPTER 5

1. Steele received considerable criticism from many citizens for his decision to replace Mc-Nelly, although it was the only rational decision to make. He provided a lengthy explanation for his decision to the press, and it was printed in the *Galveston Daily News* of February 6, 1877, and the *San Antonio Daily Express* of February 9, 1877. In it he does not discuss why he chose Hall over Armstrong, however. See Thomas W. Cutrer, "Jesse Leigh Hall," in Tyler, *New Handbook,* vol. 3, 414.

2. Durham, *Strip,* 158. Durham's memory was playing tricks in this case, as Robinson had been killed in April of 1876 while on leave to his home state of Virginia. He could not have been considered for the position.

3. Pleasanton was the county seat of Atascosa County until 1910. Today Jourdanton is the county seat, although Pleasanton is the larger of the two cities.

4. Steele, "Results," 8.

5. Oliver S. Watson was mustered into McNelly's command as a private on August 6, 1876, and honorably discharged February 1, 1877. He then was mustered into Hall's Special State Troops on January 25, 1877, with the rank of sergeant, and served through August 31, 1877. By 1880 he was working as a drayman for hire in Corpus Christi and was living in Rachel Byington's boardinghouse. The Mississippi native was born about 1850 (Texas Ranger service record, Texas State Archives; Texas Census, Nueces County, enumerated June 2, 1880, by Joseph Fitz-Simmons, 2). At this stage of his career Alfred Allee's only known criminal action was the killing of James Word during a disturbance at a New Year's dance in an obscure little community in Karnes County called Polecat, near the Goliad County line. This was the first man he killed, so it is surprising that he is so early branded a "criminal."

6. Hall to Steele, written from Goliad, February 8, 1877 (Adjutant General Papers).

7. Armstrong to Steele, "Report of Operations," February 22, 1877, written from Pleasanton (Adjutant General Papers).

8. Hall to Steele, March 2, 1877, written from Clinton, DeWitt County (Adjutant General Papers).

9. *Galveston Daily News,* March 3, 1877, reprinting correspondence from the *San Antonio Daily Herald* reporting news from Dimmit County.

10. Theodore Terry to the Hon. H. H. Boone, attorney general, State of Texas, March 9, 1877 (Adjutant General Papers; original emphasis).

11. *Galveston Daily News,* March 23, 1877. William Lawrence Rudd was one of the few

Texas Rangers who were foreign born, in his case, in Everingham, Yorkshire, England, on August 10, 1845. By 1872 he had migrated to America, landing in Galveston. In 1874 he joined McNelly's company. He served under McNelly, Hall, and Thomas Oglesby, Hall's successor. In 1883 he married Miss Evelyn Harper. Rudd was appointed sheriff of Karnes County on September 9, 1886, elected on November 2, 1886, and served until November 6, 1888 (*Yorktown News,* December 7, 1939, and January 16, 1941; Tise, *Texas County Sheriffs,* 296). He died on January 9, 1941, and is buried in the Westside Cemetery in Yorktown, DeWitt County, beside his wife, who died December 1, 1939.

12. *Dallas Daily Herald,* April 12, 1877, reporting news from Wilson County.

13. Armstrong to Steele, April 13, 1877, written at Clinton, DeWitt County. Here Armstrong signs his three-page letter as "2nd Lieut. Special State Troops."

14. Ibid.

15. *Western Chronicle,* April 27, 1877.

16. "Report of Operations," J. B. Armstrong to Adj. Gen. Steele, May 4, 1877, letter written from Pleasanton, Atascosa County.

17. Ibid.; *San Antonio Daily Express,* May 15, 1877, citing a dispatch from Eagle Pass dated May 14.

18. In a separate communication to Steele, prepared on May 14, Hall indicates that he had arrived at Eagle Pass the previous night. He points out that Armstrong already had six prisoners in jail, among them one Murray, who had broken out of jail in Pleasanton, and Joe Horner's partner in a recent stagecoach robbery (Hall to Steele, May 14, 1877, Adjutant General Papers). Horner was developing a reputation as a real desperado by this time. On April 20 word came from Uvalde that he had been captured, not by Hall's Rangers but by "members of the new company of Minute Men just being organized in Frio Canon." Their names were J. J. H. Patterson, Henry Patterson, W. B. Nicholls, Tom Leakey, and John Collins. Horner and his companion fought as long as they could. Horner was taken but his companion "made his escape amid a perfect shower of bullets." Horner was wanted for his role in the Comanche bank robbery as well as stagecoach robberies (*San Antonio Daily Express,* April 21, 1877, "Special Dispatch" from Uvalde, Uvalde County, dated April 20). The "Kimble County Roundup" is discussed briefly in Wilkins, *Law,* 126–29, and Utley, *Lone Star Justice,* 178–80.

19. J. L. Hall, "Lt. Cmdg Spec State Troops," May 16, 1877, written at Eagle Pass.

20. Ibid.

21. *San Antonio Daily Express,* May 15, 1877. Stagecoach and bank robber Joe Horner, who years later became known as Frank Canton and was a highly respected lawman in the Indian Territory and later the State of Oklahoma, is the subject of Robert K. DeArment's biography, *Alias Frank Canton.*

22. *San Antonio Daily Express,* May 22, 1877, citing a report from Uvalde dated May 21.

23. Ibid., May 24, 1877, citing a report from Castroville.

24. Ibid., May 25, 1877.

25. Hall to Steele, May 16, 1877, written at Eagle Pass (Adjutant General Papers).

26. Ibid.

27. This was Col. William Rufus "Pecos Bill" Shafter. In 1875 he gained considerable recognition for his expedition from Fort Concho across the Llano Estacado to the Pecos River in New Mexico Territory and return. This involved considerable hardship, the exploration of new

country, and some engagements with Indians. In 1876 he led another expedition against Lipan and Kickapoo Indians who lived in Mexico but who consistently raided into Texas. Shafter survived his Indian-fighting days and died in Bakersfield, California, on November 12, 1906 (Thrapp, *Encyclopedia,* 1291).

28. Hall to Steele, May 16, 1877, written at Eagle Pass (Adjutant General Papers).

Chapter 6

1. Telegram, Hall to Steele, May 29, 1877, sent from Goliad via Victoria to Austin (Adjutant General Papers).

2. *Victoria Advocate,* June 2 and 16, 1877. Newspapers frequently reported on the progress of Armstrong's wound, much more so than Watson's. For example, the *Western Chronicle* of Sutherland Springs in Wilson County, which had only recently commenced publication, reprinted in the July 6 issue an item from the *Goliad Guard* that stated that Armstrong was "fast recovering from his recent wound." The *Cuero Daily Bulletin* reported that Armstrong was "able to limp along without his crutches. The Sergeant's many friends in this section will receive this information with delight" (*San Antonio Daily Express,* August 15, citing an item from an undated *Cuero Daily Bulletin*).

3. *Western Chronicle,* June 8, 1877.

4. Hall to Steele, May 31, 1877, writing from Cuero, DeWitt County (Adjutant General Papers). Watson was discharged from the service on October 31, 1877, at Cuero. In spite of Hall's praise for the man, he remains merely a name in Ranger records with no action bringing him special recognition.

5. *San Antonio Daily Express,* August 15, 1877, citing a report from an undated *Cuero Bulletin.*

6. Governors Richard Coke and Richard B. Hubbard Executive Record Books, Texas Secretary of State, ARIS/TSLAC, reel #3481, 291.

7. Sen. J. D. Stephens from Comanche introduced a joint resolution authorizing the governor to offer the $4,000 reward. This was read on January 14 and 15, 1877, and approved on January 20 (State of Texas, *Journal,* 107, 112). Joint Resolution No. 47 authorized the governor to offer a reward of "$4000 for the arrest . . . and making appropriations therefore" (State of Texas, *Journal,* 110). In its final form, and renumbered Joint Resolution No. 1, it called for the "apprehension and delivery of the body of the notorious murderer, John Wesley Hardin, delivered within the jail house door of Travis county, to be paid out of any moneys in the treasury not otherwise appropriated" (State of Texas, *General Laws,* 189).

8. The writing on Hardin is voluminous. For his own version of his adventures, see *The Life of John Wesley Hardin, As Written by Himself,* posthumously published in Seguin, Texas, in 1896, the year following his death and reprinted most recently in 1961. I have used the original 1896 edition for any quotations, as the reprints have not remained faithful to the original. The University of Oklahoma Press edition has minor changes, hence my use of the original. Lewis Nordyke presents an acceptable biography, *John Wesley Hardin: Texas Gunman.* More recently two fine biographies have appeared: *The Last Gunfighter: John Wesley Hardin,* by the late Dr. Richard C. Marohn; and *John Wesley Hardin: Dark Angel of Texas,* by Leon Metz. Shorter works dealing with Hardin's capture include Parsons, *The Capture of John Wesley Hardin,* and idem, *Bowen and Hardin. Bowen and Hardin* has been updated by Bowden and Cummins, *Texas Desperado in Florida: The Capture of Outlaw John Wesley Hardin in Pensacola, 1877.* For a

definitive biography of Armstrong's partner in Hardin's capture—John R. Duncan—see Miller, *Bounty Hunter.*

9. Nordyke, *John Wesley Hardin,* 190.

10. Joshua Bowen to Jane Swain [Hardin], May 5, 1877. Joshua Bowen was Jane Bowen Hardin's uncle (John Wesley Hardin Letters).

11. Duncan's service record shows he "rangered" from July 15 through August 31, 1877. He earned $40 per month, for a total of $60, according to Hall's August 31 muster and payroll. Curiously, another muster and payroll, dated November 30, 1877, at Corpus Christi, shows he received $100 for service as a private from August 30 through November 15.

12. J. H. Swain to Jane Hardin, August 25, 1877. This communication is written on the letterhead of attorneys-at-law J. S. Clark and David P. Lewis of Decatur, Alabama (John Wesley Hardin Letters).

13. *Daily Democratic Statesman,* August 18, 1877, "Hotel Arrivals" column.

14. Ibid., August 19, 1877, "Personal and Local Dots" column.

15. Ibid., August 29, 1877. This is from a lengthy interview with Armstrong by a *Daily Democratic Statesman* reporter.

16. J. H. Swain to Jane Hardin, August 25, 1877 (John Wesley Hardin Letters).

17. Austin *Daily Democratic Statesman,* August 29, 1877.

18. Ibid. The telegrams from Armstrong are now preserved in the Adjutant General Papers.

19. T. R. Armstrong, "Capture," 1–2. The accompanying letter from Tom Armstrong is dated July 30, 1934, and thus was written almost fifty-nine years after the capture. He writes, "I have heard some other accounts, including one that was attributed to Hardin himself and published by a newspaper as being his memoirs written while in the penitentiary. All of these vary widely from the account I heard from Papa on two occasions. He told me the story once in 1905, when I stayed out of school to spend the winter with him on the ranch after John was killed [on May 6, 1905, when he was thrown from a horse]. During that time, he also recounted a number of his other experiences while with McNelly's Rangers. The only other time I heard him tell the story was in 1910 when I went to the ranch to spend the Christmas holidays with him and he told the story to my two classmates, Harry Caesar and Skip Simpson, and me."

20. Ibid.

21. Armstrong later explained that Hardin's six-shooter was not in a traditional holster but stuck inside his pants with one of his suspender straps through the trigger guard; he was wearing it in this manner because of an old wound. T. U. Taylor writes that he and Armstrong discussed the capture while traveling from Texarkana to Austin "something like fifteen years ago," which would date it about 1910 (Taylor, "New Light," 11, 16).

22. T. R. Armstrong, "Capture," 4.

23. Ibid.

24. *Montgomery Advertiser and Mail,* August 28, 1877.

25. T. R. Armstrong, "Capture," 5.

26. *Galveston Daily News,* August 23, 1895, in an interview with Duncan at the time of the killing of Hardin, which occurred in El Paso on August 19, 1895.

27. Hardin, *Life,* 120.

28. Hardin, *Life,* 121; *Daily Democratic Statesman,* August 29, 1877, editorial, "Casting Out of Devils—How Texas Does It."

29. W. D. Chipley letter to Hubbard, August 28, 1877, Texas State Archives.

30. *Montgomery Advertiser and Mail*, September 19, 1877; *Mobile Register*, September 22, 1877; Records of the Governor (Richard B. Hubbard), letter to Secretary of State Isham G. Searcy from Governor Hubbard, September 4, 1877; W. D. Chipley letter to Hubbard, August 28, 1877, all in Texas State Archives.

31. He was pardoned after sixteen years in prison, after numerous influential citizens asked Governor Hogg to do so. See Crouch, "'That Good Citizens Ask It.'"

32. Miller, *Bounty Hunter,* 222–23.

33. *El Paso Daily Times,* September 8, 1895, reprint of an article from the *Dallas News,* September 6, 1895.

34. *Pensacola News,* January 14, 1911.

35. Williamson, "William Dudley Chipley."

36. Jennings, *Texas Ranger,* 154–55. McKinney had nothing at all to do with the capture, so Jennings's memory is faulty. He probably meant to write that the five were Duncan, Armstrong, Chipley, Deputy Perdue, and Sheriff Hutchinson. He goes on to say that "each one" got $800 for the capture. There were five of them, in his mind, and the $4,000 reward divided by five would give each one $800. Perhaps he confused the name "Chipley," which he does not mention, with "Charley McKinney," whom he knew.

37. *San Antonio Daily Express,* May 3, 1913.

38. *The Lawless Breed,* produced by Raoul Walsh, with Rock Hudson starring as Hardin, was released in 1952 by Universal Pictures. A motion picture was released in 2001 by Price-Greisman Productions entitled *Texas Rangers,* ostensibly based on Durham's *Taming the Nueces Strip.* It also features Armstrong, played by Robert Patrick, in a minor role.

CHAPTER 7

1. *Dallas Daily Times Herald,* August 24, 1895.

2. *Atlanta Daily Constitution,* September 1, 1877. This is a lengthy article based on an interview with Duncan and headlined "A Texas Outlaw," with subheadline, "Arrest of Jack Swayne —A Noted Bandit Who Has Killed Twenty-Seven Men." The interview is datelined Pensacola Junction, August 25.

3. Ranger service record, J. B. Armstrong, Texas State Archives.

4. *Daily Democratic Statesman,* August 26, 1877.

5. Supt. Thomas J. Goree to Steele, written from Huntsville, September 1, 1877, Adjutant General Papers.

6. *Weatherford Exponent,* September 22, 1877.

7. *Galveston Daily News,* October 25, 1877.

8. Ibid., December 23, 1877, printing a letter from Cuero dated December 18, 1877.

9. Ibid., December 26, 1877, written in Cuero on December 21. The complete letter from "Total Wreck" appears in Parsons, "Rangers."

10. Henry Hutchings, adjutant general and chief of staff, General Orders No. 22, Austin, May 2, 1915, Tobin Armstrong Papers. This is a résumé of Armstrong's rank from a private in McNelly's company through his position as lieutenant colonel and assistant chief of ordnance. It includes an announcement of Armstrong's death.

11. *Daily Democratic Statesman,* January 5, 1878.

12. *Dallas Daily Herald,* February 22, 1878, citing a news item from an undated *Victoria Advocate.*

13. Pattie Townes was later a schoolteacher in Travis County (Texas Census, Travis County, enumerated June 5–7, 1880, by J. S. Churchill, 200). Thomas Devine, formerly a Ranger companion of Armstrong's, had been discharged from the Rangers on December 1, 1876. Ella Mabry was the daughter of noted cattleman Seth Mabry. She was still a teenager at the time of the wedding ("Schedule," 112). Florence Jackson and her brother Alex M. Jr. both attended the ceremony. At the time of the marriage she was about nineteen and he, twenty-five. They were the children of C. C. and A. M. Jackson, a prominent Austin attorney ("Schedule," 109). Sterling Fontaine Grimes, born on December 12, 1842, in Lincoln County, Kentucky, had served as a Confederate soldier, at one time with Gen. John Hunt Morgan in the Indiana and Ohio raid, but he was captured in July of 1863. He spent the remainder of the war as a Union prisoner at Camp Douglas, Illinois, and was exchanged in March of 1865. After the war he settled in Clinton, DeWitt County. After his admission to the bar in 1871 he began a long legal career. In time, he partnered in the firm of Crain, Kleberg & Grimes of Cuero. During his career he prosecuted various members of the Sutton-Taylor families, including John Wesley Hardin. Grimes died in Cuero on May 4, 1918 (Huson, *District Judges,* 105–107; F. W. Johnson, *History,* vol. 5, 2489–90). Robert J. Kleberg had received superior education in private schools prior to obtaining a law degree from the University of Virginia in 1880. After the death of Richard King he assumed management of the King Ranch. Julia M. Pease was the daughter of former governor Elisha M. Pease. In 1870 she was sixteen years old, her father now practicing law (Texas Census, Travis County, enumerated July 30, 1870, by O. T. Zink, 205). Ella R. Carter would, on December 12, 1883, marry Dudley Goodall Wooten, a graduate of the law school of the University of Virginia. Their union produced two children, but Mrs. Wooten died on February 9, 1886 (*Austin Daily Statesman,* February 12, 1886; Tyler, "Dudley Goodall Wooten," *New Handbook,* vol. 6, 1074; "Scrapbook of Newspaper Clippings, Poems, Engravings, and Other Material," in Hilgartner-Palm Papers, Austin History Center).

14. *Daily Democratic Statesman,* February 21, 1878.

15. Ibid., February 22, 1878.

16. *Galveston Daily News,* February 21 and 26, 1878. The list of guests mentioned in the *Galveston Daily News* but not in the *Daily Democratic Statesman's* reports includes Mr. and Mrs. R. H. Ward. She was the former Miss Annie DeCordova, who, along with several other Austin ladies—Miss Mollie Durst, Miss Phoebe Peck, Miss Clara Haralson, and Miss Salome Smith—had given a "magnificent supper" to the Travis Rifles in November of 1874. At the Ward-Cordova wedding on November 17, 1875, Mollie Durst and Clara Haralson were the bride's attendants.

17. *Daily Democratic Statesman,* February 26, 1878.

18. *Dallas Weekly Herald,* April 6, 1878.

19. *Daily Democratic Statesman,* April 5, 1878.

20. *San Antonio Daily Herald,* April 13, 1878.

21. *Daily Democratic Statesman,* May 31, 1878.

22. Ibid., June 30, 1878.

23. Miller, *Sam Bass,* 246–55.

24. *Galveston Daily News,* July 24, 1878.

25. *San Antonio Daily Express,* September 12, 1895.

26. *Daily Democratic Statesman,* November 15, 1878.

27. Ibid., October 2, 1879.

CHAPTER 8

1. Texas Census, Travis County, enumerated June 3, 1880, by Thomas A. Taylor, 252. The 1881–82 Austin city directory shows Armstrong as a resident on the north side of Hickory Street between Brazos and San Jacinto (*Morrison & Fourmey's, for 1881–82,* 48), and the boarding-house of Mrs. M. J. Durst, widow, at the same address (77). The date of Maria Josephine Armstrong's birth is from Smith, *Armstrong Chronicle,* 124.

2. *Edwards & Church's General Directory of the City of Austin for 1883–84* shows Armstrong as a clerk (45). The C. R. Johns & Joseph Spence firm notice is found on page 97.

3. In the Tobin Armstrong Collection.

4. *Morrison & Fourmey's, for 1881–82,* 48, 77; ibid., for years 1900–01, 48.

5. *Austin Daily Statesman,* June 4, 1882.

6. The Austin Greys, Company A, Second Regiment Texas Volunteer Guard, was organized July 4, 1876. R. P. Smyth was elected captain. It began with a membership of forty-five young men (*Edwards & Church's,* 38).

7. Will Lambert (1840–98) had a distinguished career in military, journalism, and legislative circles. He was orphaned in 1848, and in 1851 started as a printer's apprentice on the staff of the *San Antonio Weekly Ledger* that same year. By 1857 he was in Austin working on a newspaper; in 1859 he served with Edward Burleson Jr.'s Ranger company. During the Civil War he served under Col. Henry McCulloch in the Texas Mounted Rifles, was wounded, and then paroled in 1865. In 1868 he was publisher of the *Houston Telegraph;* in 1877, editor of the *Houston Age.* He was commissioned a colonel in the Travis State Guard. In January 1881 he was appointed acting secretary of the Texas House of Representatives. He died in Houston in October 1898 (Thomas W. Cutrer, "Will Lambert," in Tyler, *New Handbook,* vol. 4, 45–46).

8. *Austin Daily Statesman,* August 30, 1882.

9. "Durst and Armstrong Families," [1–2]. See also Carolyn Hyman, "Joseph Durst," in Tyler, *New Handbook,* vol. 2, 738.

10. "Durst and Armstrong Families," [2].

11. Ibid., [3–4]. See also Smith, *Armstrong Chronicle,* 59–61.

12. "Durst and Armstrong Families," [4–6]. See also Smith, *Armstrong Chronicle,* 125–27.

13. Mifflin Kenedy, born in Pennsylvania in 1818, was the son of Quaker parents. His early years were spent teaching. After a voyage as a cabin boy to Calcutta, he decided to make a career of river navigation. From 1836 to 1842 he was clerk and acting captain on steamers on the Ohio, Missouri, and Mississippi rivers. He met Richard King in Florida, with whom he entered into a steamship partnership in 1850, M. Kenedy & Company. In 1860 Kenedy and King bought into the Santa Gertrudis Ranch as full partners. In 1868 Kenedy sold his share of the ranch and purchased the Laureles Ranch near Corpus Christi. The steamship company profited during the Civil War, and the pair ended their partnership in 1874. Kenedy was among the first ranchers to fence his property. In 1876 he entered the new field of railroad construction

to help Uriah Lott build the Corpus Christi, San Diego & Rio Grande Railroad from Corpus Christi to Laredo. He died at Corpus Christi on March 14, 1895 (John Ashton, "Mifflin Kenedy," in Tyler, *New Handbook,* vol. 3, 1064–65).

14. An idea succinctly expressed by his grandson, Tobin Armstrong, in a telephone interview, September 26, 2005.

15. Bethel Coopwood was much more than a superior lawyer. He was born in 1829 in Alabama and came to Texas in 1846 to fight in the Mexican War. In 1854 he went to California, where he was admitted to the bar. He returned to Texas, where he was soon recognized as an "able lawyer and Spanish scholar in the lower Rio Grande valley." He served in the Confederate Army during the Civil War. Later he contributed articles and book reviews to the Texas State Historical Association *Quarterly.* He died in Austin on December 26, 1907 (Margery H. Krieger, "Bethel Coopwood," in Tyler, *New Handbook,* vol. 2, 316). The 1880 census shows some of his travels. He was born in Alabama; his wife, Josephine, was born in New Jersey. Bethel Jr. was born in California; son William was born in New Mexico; James D., Mary E., and Emma were born in Texas; and L. D. was born in Mexico (Texas Census, Travis County, enumerated June 7, 1880, by J. H. Burns, 237).

16. Armstrong to Kenedy, July 18, 1883, copy in Tobin Armstrong Collection.

17. Armstrong to Kenedy, February 15, 1884, copy in Tobin Armstrong Collection.

18. Ibid.

19. Armstrong to Kenedy, March 21, 1884, copy in Tobin Armstrong Collection.

20. Mariah Susannah Ready Armstrong later returned to McMinnville, Tennessee, to live with her daughter Betavia Armstrong Beech. She died there on November 9, 1885 (Smith, *Armstrong Chronicle,* 127).

21. The "precinct" not yet heard from, as Armstrong describes his wife's pregnancy, was born July 7, 1884, and named John Barclay (not Barkley). He was their third child. Maria Josephine was born April 5, 1879; Jamie Durst, January 5, 1881; Charles Mitchell (their fourth child), November 8, 1886; Julia Katherine, February 5, 1889; Elliott Ropes, October 9, 1890; and Thomas Reeves, September 5, 1892. Birthdates from Smith, *Armstrong Chronicle.*

22. Armstrong to Capt. Mifflin Kenedy, June 25, 1884, copy in Tobin Armstrong Collection.

23. Maria Josephine lived until 1972; Jamie Durst lived until 1963; John Barclay lived until 1905; Charles Mitchell lived until September 13, 1941; Julia Katherine lived until December 26, 1991; Elliott Ropes lived until May 31, 1898; and Thomas Reeves lived until March 3, 1896.

24. "Durst and Armstrong Families," [7].

25. Ibid., [7]. In 1925 the Stewarts purchased the renowned Oak Alley Plantation near Vacherie, Louisiana, once one of the most beautiful antebellum homes in the southern states. At the time it and the outbuildings were nearly in ruins, but the Stewarts restored them and then lived in the mansion. When they passed on they were buried on the grounds (telephone interview with Tobin Armstrong, December 6, 2003; also see Smith, *Armstrong Chronicle,* 170–71).

26. "Durst and Armstrong Families," [8].

CHAPTER 9

1. Smith, *Armstrong Chronicle,* 148–49.

2. Robert J. Kleberg to Adj. Gen. W. H. King, May 29, 1888, Adjutant General Papers.

3. Ibid.

4. Gov. James S. Hogg, Special Order No. 6, 259 (Texas State Archives), requirements for applying to become a Special Ranger. Although the order is dated March 6, 1891, the requirements were certainly the same when Armstrong rejoined the Ranger force in 1888.

5. "Descriptive List" document in Armstrong's Service Record file, Texas State Archives.

6. Stephens, *Sketches*, 65–69.

7. John Chenneville, a Louisiana native, had served with the Austin police for years and thus had considerable experience dealing with rough men. Paulin S. Coy began his Ranger career under Thomas L. Oglesby and the Special Force, formerly McNelly and Hall's Special State Troops, mustering in on December 1, 1879, and serving until November 30, 1880. He then continued under Oglesby (who now commanded Company F of the Frontier Battalion) from September 1 until November 30, 1881. He asked to become a Special Ranger at King's Santa Gertrudis Ranch on March 12, 1892. The thirty-five-year-old San Antonio native gave his occupation then as "Stockman" (Coy's Ranger service records, Texas State Archives). G. B. Greer applied for reenlistment on March 27, 1891, while living in Harris, Edwards County. He was thirty-eight years old, a native of Warrensburg, Missouri, and a "Wool grower" by occupation (G. B. Greer's Ranger service records, Texas State Archives). W. J. Greer served under L.P. Sieker in Company D from July 25, 1885, to October 5, 1885 (W. J. Greer's Ranger service records, Texas State Archives). Augustine Montague "Gus" Gildea was probably born on April 11, 1854, in San Antonio, although his service record shows he was born in DeWitt County. When he enlisted in Austin on August 23, 1897, he was forty-three years old and gave his occupation as farmer. He enlisted on December 5, 1890, in Company D. The record shows he served under Capt. Frank Jones from June 1, 1887, until November 30, 1890. He was a cowboy in the late 1860s on his father's ranch in Live Oak County, Texas, and by the late 1870s was working in the New Mexico and Arizona territories. His Special Ranger service was from June 1, 1887, until November 30, 1890, in La Salle, Presidio, Duval, and Brewster counties. Gildea died on August 10, 1935, in Douglas, Arizona (Rasch, "'Gus Gildea,'" 1–7). E. R. Jenson was a native of Junen, Denmark, thirty-eight years old when he requested appointment as a Special Ranger at Alice, Duval County, on February 9, 1892. He gave his occupation as "Ranchero." He served under McNelly from August 1, 1876, until February 1, 1877, and also under Hall, from January 25, 1877, until his honorable discharge dated October 15, 1878 (Jenson's Ranger service records, Texas State Archives). John G. Kenedy served as a Special Ranger in Capt. J. A. Brooks's company from 1888 to 1896. His reports to Adj. Gen. W. H. Mabry on Kenedy Pasture Company letterhead, listing Mifflin Kenedy as president and himself as secretary, consistently state, "Nothing to report." A single card in his service file states that he was born in Brownsville, Cameron County, on April 22, 1856. Samuel R. Pickett remains an elusive figure in Ranger history. He was born circa 1865 and enlisted in Capt. George H. Schmitt's Company C in Pearsall, Frio County, on September 1, 1887. He served in North, South, and Central Texas and was discharged on November 30, 1887, a full two-year Ranger career. As a Special Ranger he was employed on the King Ranch, but by the 1890s was running cattle in La Salle County. He died on December 23, 1903, near San Antonio, with no family and virtually in poverty (Stephens, *Sketches*, 118–20). J. E. Van Riper was born in Stockton, Johnson County, about 1853. He first served under Lt. G. W. Campbell of Company B from September 1, 1877, until May 31, 1878, then under Capt. June Peak from September 1 to December 31, 1878. His service record shows his occupation as U.S. deputy marshal and that he "enlisted at Austin as a special assistant to

the Rangers, in his capacity as a U.S. official" (Van Riper's Ranger service records, Texas State Archives). No document in Ernest Rogers's file verifies any service as a Special Ranger. He did serve under Capt. Frank Jones as a regular Ranger from September 1, 1887, until April 1, 1889 (Rogers's Ranger service records, Texas State Archives).

8. "John Barclay Armstrong," in Webb, *Handbook,* vol. 1, 69; Thomas W. Cutrer, "John Barclay Armstrong," in Tyler, *New Handbook,* vol. 1, 244; Ferris E. Stovel, Judicial and Fiscal Branch, Civil Archives Division, National Archives, correspondence with author, May 15, 1978; F. W. Johnson, *History,* vol. 3, 1359.

9. George J. Reynolds was born in England about 1855. He first served under Capt. John C. Sparks in Company C from September 9 until November 30, 1877, then enlisted under Capt. Frank Jones as a Special Ranger on June 27, 1889, at Uvalde, Uvalde County. A later request for appointment is dated March 7, 1892; Capt. J. S. McNeel wrote on the application, "I take great pleasure in recommending Mr. Reynolds" (Reynolds's Ranger service records, Texas State Archives).

10. Private Broome was forty-six years old when he enlisted on April 13, 1884, in Company D at Austin. An earlier request to be appointed, dated March 13, 1881, shows he was born at Utica, Mississippi, and worked as a U.S. deputy marshal (Ranger service records, Texas State Archives). I have been unable to find further information on J. W. Mathers.

11. Harris and Sadler, *Texas Rangers,* 57.

12. Utley, *Lone Star Justice,* 294. Charles M. Robinson III devotes an entire chapter to discussion of some of the excesses of "Ranger justice" meted out to innocent Mexicans along the border (*Men,* 264–80). Don Graham reviews Webb's *The Texas Rangers* and writes that Webb and others lionized the Rangers; their work "whitewashes their [Ranger] excesses, offering excuses and justifications for illegal actions such as torture and murder." Graham accuses Webb of being an "adept apologist for their conduct" and quotes him as saying, "Affairs on the border cannot be judged by standards that hold elsewhere" ("Fallen Heroes," 70). In spite of the recent revisionist publications dealing with the Texas Rangers, no one has discovered any type of mistreatment of prisoners, Anglo or Mexican, performed by Armstrong. Ironically, before and during the invasion of Mexico in 1875, the mistreatment of prisoners by a McNelly's Ranger who was either born in Mexico or was of Mexican parents—Jesús Sandoval—can be documented.

13. There are three colored broadsides advertising the semicentennial event in the Adjutant General Papers; the note concerning the "grand sham battle" appears in Circular No. 3.

14. *Galveston Daily News,* June 16, 1889.

15. Armstrong to Adj. Gen. Woodford Haywood Mabry, March 6, 1891, Adjutant General Papers.

16. Oath dated April 28, 1891, in Armstrong's Ranger service file, Texas State Archives.

17. Special Orders A.G.O., No. 68, dated July 19, 1892, 295, Texas State Archives.

18. Special Orders A.G.O., No. 182, dated June 24, 1892, 295, Texas State Archives.

19. Oath dated April 8, 1893, Adjutant General Papers.

20. Oath dated June 26, 1895, Adjutant General Papers.

21. Samuel D. DeCordova was the son of Phineas D. and J. D. DeCordova and was born about 1861. In 1900 he was in partnership with his father in the firm of DeCordova & Son, with offices at 704 Congress Avenue (*Morrison & Fourmy's, 1900–01,* 86).

22. Oath dated January 12, 1900, Adjutant General Papers.

23. Statement, February 7, 1902, Adjutant General Papers.

24. The information on Armstrong as a pioneer and rancher is primarily from interviews with the late Tobin Armstrong, esp. the telephone interview of September 26, 2005.

25. "The Armstrong Ranch," [20–21]. According to the records of Oakwood Cemetery, Austin, Mollie succumbed to Bright's disease, any of several diseases of the kidneys. The attending physician was Dr. Frank P. McLaughlin, whose office was on the corner of West 7th Street and Congress Avenue. Apparently, there was a social stigma connected with the malady of rabies, thus it was given out that she died of Bright's disease. The original cemetery record is housed in the Austin History Center; the entry for Mary Josephine Durst is found on 142; for Mollie Durst Armstrong, on 292; and for John Barkley Armstrong, on 150. The scratch which claimed the life of Mrs. Armstrong must have been very early in December, as, typically, the incubation period for rabies is three to seven weeks. After the incubation period is over "a tingling sensation usually develops at the site of the animal bite [or scratch]; a more generalized skin sensitivity may occur and changes in temperature become very uncomfortable. As the virus spreads, foaming at the mouth may occur. . . . Uncontrolled irritability and confusion may follow, alternating with periods of calm" (David E. Larson, M.D., editor-in-chief, *Mayo Clinic Family Health Book* [New York: William Morrow and Company, 1996], 395).

26. *Austin Daily Statesman,* December 28, 1897.

27. Henrietta Maria Morse Chamberlain was born on June 21, 1832, in Boonville, Missouri. On December 10, 1854, she married Richard King, who predeceased her on April 14, 1885. With her education and sound common sense it was perhaps natural for her to become the supervisor of housing and education for the families of the rancheros working on the King Ranch. She offered 75,000 acres of right-of-way to Uriah Lott and B. F. Yoakum for the construction of the St. Louis, Brownsville & Mexico Railway. In 1904 she furnished land for the town sites of Kingsville and Raymondville, located on the railway. Mrs. King died on March 31, 1925, and is buried in Kingsville (Edgar P. Sneed, "Henrietta Chamberlain King," in Tyler, *New Handbook,* vol. 3, 1106; Bruce S. Cheesman, "Richard King," in Tyler, *New Handbook,* vol. 3, 1107–08). John G. Kenedy was born about 1852 and is found in the 1880 Nueces County census with occupation listed as sheep raiser. His father is shown as sixty-one-year-old Mifflin Kenedy, a "Ranchero," and his mother, Petra Vidal, as age fifty-two. Enumerator Cornelius Cahill added the following at the end of the enumeration for Kenedy's ranch: "Here ends the live stock Rancho de los 'Laureles' containing 172,000 acres under fence, Mifflin Kenedy owns and [is the] proprietor who conducts the same in person." Cahill had enumerated seventy individuals working for Mifflin Kenedy (Texas Census, Nueces County, enumerated June 29, 1880, 46–47).

28. Lea, *King Ranch,* vol. 2, 541.

29. Ibid., 544. Also see the *Austin Statesman,* July 6, 1904, citing a special from Brownsville, and the *Galveston Daily News* of July 5 and 6, 1904.

30. Information on the various towns is from Allhands, *Gringo Builders,* 81–102.

31. Smith, *Armstrong Chronicle,* 155.

32. *Austin Daily Statesman,* September 11, 1912.

33. Ibid. Memorial message on separate headstone in the Armstrong family plot.

34. Armstrong's standard certificate of death (#11310). Additional information on the death certificate, his full name, date of birth, and parents' names, was provided by his son Charles M. Armstrong.

35. *Cuero Daily Record,* May 6, 1913. Many newspapers throughout Texas reported Armstrong's death.

36. *San Antonio Daily Express,* May 2, 1913; *Austin Statesman,* May 2, 1913.

37. *Austin Statesman,* May 3, 1913. The Armstrong plot is in section 2, lot #777. The monument is nearly twenty feet tall, with inscriptions for John B. on one side, Mary H. and Elliott Ropes on the next side, John B. Jr. and Charles M. on the third side, and, on the fourth side, Charles M. Armstrong Jr., who "Died in Service of His Country/1922–1943." There is a separate headstone for Mary Josephine Atwood Durst.

38. *Austin Statesman,* May 2, 1913.

39. General Orders No. 22, Texas State Archives.

40. *San Antonio Daily Express,* May 2, 1913. This newspaper provided editorial comments on Armstrong's life following the special telegram from Kingsville reporting his death.

41. *San Antonio Daily Light,* January 1, 1913.

42. Smith, *Armstrong Chronicle,* 174.

43. *San Antonio Daily Light,* January 24, 1918; *Austin Statesman,* September 15, 1941.

44. John M. Bennett, letter of April 13, 1978

45. Ibid.

Selected Bibliography

ARCHIVES AND MANUSCRIPT SOURCES

Adjutant General Papers. Muster and payrolls, correspondence to and from the adjutant general, special orders, other relevant documents. Adj. Gen. Dept. Archives & Information Services. Division of the Texas State Library & Archives Commission (hereafter ARIS/TSLAC), Austin.

Armstrong, Lavanda M. "Genealogy and Family Collections, Etc. of the Armstrong-Ready Family of Tennessee." May 23, 1933. Tobin Armstrong Collection, Armstrong Ranch.

Armstrong, Thomas R. "Capture of John Wesley Hardin by John B. Armstrong of the Texas Rangers." Typescript prepared by T. R. Armstrong at the request of his brother Charles M. Armstrong. Copy in author's possession.

Armstrong family bible. Tobin Armstrong Collection, Armstrong Ranch.

"The Armstrong Ranch: History of a Noble Pioneer Family of Texas." No author. Typescript. Tobin Armstrong Collection, Armstrong Ranch.

Aten, Ira. "Six and One Half Years in the Ranger Service Fifty Years Ago." Typescript. Center for American History, University of Texas, Austin.

Brown, Frank. "Annals of Travis County and the City of Austin." Typescript. Austin History Center, Austin, Texas. The prospectus dates publication as 1901.

Callicott, William C. Correspondence. Walter Prescott Webb Papers. Center for American History, University of Texas, Austin.

"Cemetery Record." Austin History Center, Austin, Texas.

"Delegation from Dalhart Big Part of Gathering." Roy W. Aldrich Newspaper Clippings Collection, folder no. 201. Sul Ross State University, Alpine, Texas.

"The Durst and Armstrong Families." No author, no date. Tobin Armstrong Collection, Armstrong Ranch.

John Wesley Hardin Letters. Special Collections, Albert B. Alkek Library, Texas State University, San Marcos.

Hilgartner-Palm Papers. Austin History Center, Austin, Texas.

Hubbard, Richard B. Correspondence. ARIS/TSLAC, Austin, Texas.

Roberts, A. S. "Reminiscences." Typescript. Center for American History, University of Texas, Austin.

"Schedule of Inhabitants in the City of Austin, County of Travis, State of Texas, 1875." Austin History Center, Austin, Texas.

Steele, William. "Results of Operations of State Troops Since August 1, 1876 to December 31, 1881." Adj. Gen. Dept. ARIS/TSLAC, Austin, Texas.

U.S. Bureau of the Census. Census for Tennessee: Cannon County (free), 1850; Cannon County (slave), 1850; Warren County (free), 1860; Warren County (slave), 1860.

U.S. Bureau of the Census. Census for Texas: Nueces County, 1880; Travis County, 1860, 1870, 1880.

Weed, V. O. "Recollections of V. O. Weed." Center for American History, University of Texas, Austin.

NEWSPAPERS

Atlanta Daily Constitution, 1877.

Austin American, 1927, 1935.

Austin Daily Statesman, 1886.

Austin Statesman, 1897, 1904, 1913, 1935.

Colorado Citizen (Columbus), 1875.

Cuero Daily Record, 1913.

Daily Democratic Statesman (Austin), 1873, 1874, 1876, 1877, 1878, 1882.

Daily State Gazette (Austin), 1875–1878.

Dallas Daily Herald, 1877, 1878.

Dallas Daily Times Herald, 1895.

Dallas Weekly Herald, 1878.

El Paso Daily Times, 1895.

Galveston Daily News, 1875–1878, 1889, 1895.

La Grange Journal, 1921.

Mobile [Alabama] *Register,* 1877.

Montgomery [Alabama] *Advertiser and Mail,* 1877.

Pensacola [Florida] *News,* 1911.

San Antonio Daily Express, 1876, 1877, 1895, 1913, 1929.

San Antonio Daily Herald, 1875, 1878.

San Antonio Daily Light, 1914, 1918.

Victoria Advocate, 1876, 1877.

Weatherford Exponent, 1877.

Western Chronicle (Sutherland Springs), 1877.

Yorktown News, 1939, 1941.

BOOKS AND ARTICLES

Allhands, J. L. *Gringo Builders.* [Joplin, Mo., & Dallas, Tex.:] Privately printed, 1931.

———. *Railroads to the Rio Grande.* Salado, Tex.: Anson Jones Press, 1960.

———. *Uriah Lott.* San Antonio, Tex.: Naylor, 1949.

Atchison, Sharon. *Texas & Southwestern Cattle Raisers' Association Records: 1877–1968.* 1969. Rev. Anne H. Takenoto. Fort Worth: Texas Printing Co., 1972.

Barkley, Mary Star. *History of Travis County and Austin 1839–1899.* Waco, Tex.: Texian Press, 1963.

Bennett, John M. *Those Who Made It: The Story of the Men and Women of National Bank of Commerce of San Antonio, 1903–1998.* [San Antonio, Tex.:] Privately printed, 1978.

Bowden, Jesse Earle, and William S. Cummins. *Texas Desperado in Florida: The Capture of Outlaw John Wesley Hardin in Pensacola, 1877.* Pensacola, Fla.: Pensacola Historical Society, 2002.

Britton, Frank L. *Report of the Adjutant General of the State of Texas for the Year 1873.* Austin: Cardwell & Walker, 1874.

By-Laws, Rules and Regulations and Names of Members and Minutes of the Annual Meeting at El Paso, Texas, March 18–19, 1913 of the Cattle Raisers Association of Texas. Fort Worth: Texas Printing Co., 1913.

By-Laws, Rules, Regulations and Names of Members of the Cattle Raisers Association of Texas. Fort Worth: Texas Printing Co., 1899.

C. D. Morrison & Co.'s Directory of the City of Austin for 1879–80. Marshall, Tex., 1879.

Crouch, Barry. "'That Good Citizens Ask It': The Pardon of John Wesley Hardin." *Quarterly* (of the National Association for Outlaw and Lawman History) 23, no. 3 (July–September 1999): 10–23.

DeArment, Robert K. *Alias Frank Canton.* Norman: University of Oklahoma Press, 1996.

Dickey, Imogene Bentley. *Early Literary Magazines of Texas.* Austin, Tex.: Steck-Vaughn, 1970.

Durham, George. "On the Trail of 5100 Outlaws: The Inside Story of McNelly's Texas Rangers." As told to Clyde Wantland. Ed. Harry E. Maule. *West* magazine serial. N.p.: Doubleday, Doran & Co., 1934–35.

———. *Taming the Nueces Strip: The Story of McNelly's Rangers.* As Told to Clyde Wantland. 1962. Austin: University of Texas Press, 1975.

Edwards & Church's General Directory of the City of Austin for 1883–84. Austin: E. W. Swindells, 1883.

Fisher, E. Lee. *Pendencia Creek: The Life and Times of a Texas Gunfighter.* Baltimore, Md.: PublishAmerica, 2004.

Fisher, O. C., with J. C. Dykes. *King Fisher: His Life and Times.* Norman: University of Oklahoma Press, 1966.

Graham, Don. "Fallen Heroes." Book review. *Texas Monthly* (February 2005).

Haley, J. Evetts. *The XIT Ranch of Texas and the Early Days of the Llano Estacado.* Norman: University of Oklahoma Press, 1953.

Hardin, John Wesley. *The Life of John Wesley Hardin, from the Original Manuscript, As Written by Himself.* Seguin, Tex.: Smith & Moore, 1896.

———. *The Life of John Wesley Hardin, As Written by Himself.* 1896. Intro. Robert G. McCubbin. Norman: University of Oklahoma Press, 1961.

Harris, Charles H., III, and Louis R. Sadler. *The Texas Rangers and the Mexican Revolution: The Bloodiest Decade, 1910–1920.* Albuquerque: University of New Mexico Press, 2004.

Huson, Hobart. *District Judges of Refugio County.* Refugio, Tex.: Refugio Timely Remarks, 1941.

Jennings, N. A. *A Texas Ranger.* 1899. Foreword J. Frank Dobie; intro. Stephen L. Hardin. Norman: University of Oklahoma Press, 1997.

Johnson, Benjamin Heber. *Revolution in Texas: How A Forgotten Rebellion and Its Bloody Suppression Turned Mexicans into Americans.* New Haven, Conn.: Yale University Press, 2003.

Johnson, Frank W. *A History of Texas and Texans.* Ed. Eugene C. Barker. Chicago, Ill.: American Historical Society, 1916.

Kelton, Elmer. *Captain's Rangers.* New York: Bantam Books, 1981.

Lea, Tom. *The King Ranch.* Boston, Mass.: Little, Brown and Co., 1957.

Marohn, Richard C. *The Last Gunfighter: John Wesley Hardin.* College Station, Tex.: Creative Publishing Co., 1995.

Mercantile and General City Directory of Austin, Texas—1872–73. Austin, Tex.: Gray & Moore, 1872.

Metz, Leon C. *John Wesley Hardin: Dark Angel of Texas.* El Paso, Tex.: Mangan Books, 1996.

Miller, Rick. *Bounty Hunter.* College Station, Tex.: Creative Publishing Company, 1988.

———. *Sam Bass & Gang.* Austin: State House Press, 1999.

Morrison & Fourmy's General Directory of the City of Austin, Texas for 1877–78. Austin, Tex.: Eugene Von Boeckmann, 1877.

Morrison & Fourmy's General Directory of the City of Austin for 1881–82. Austin, Tex.: E. W. Swindells, 1881.

Morrison & Fourmy's General Directory of the City of Austin—1887–88. Galveston, Tex.: Morrison & Fourmy, 1887.

Morrison & Fourmy's General Directory of the City of Austin, Texas for 1900–01. Galveston, Tex.: Morrison & Fourmy, 1900.

Nordyke, Lewis. *John Wesley Hardin, Texas Gunman.* New York: William Morrow, 1957.

Parsons, Chuck. *Captain L. H. McNelly, Texas Ranger: The Life and Times of a Fighting Man.* Austin, Tex.: State House Press, 2001.

———. *The Capture of John Wesley Hardin.* College Station, Tex.: Creative Publishing Co., 1978.

———. *"Pidge" A Texas Ranger from Virginia. The Life and Letters of Lieutenant T. C. Robinson, Washington County Volunteer Militia Company "A."* Wolfe City, Tex.: Henington Publishing Co., 1985.

———. "Rangers in Lockhart." *Plum Creek Almanac* 22, no. 2 (Fall): 136–39.

———, with Marjorie Parsons. *Bowen and Hardin.* College Station, Tex.: Creative Publishing Co., 1991.

Preece, Harold. *Lone Star Man: Ira Aten, Last of the Old Time Texas Rangers.* New York: Hastings House, 1960.

Rasch, Philip J. "'Gus' Gildea—An Arizone [*sic*] Pioneer." English Westerners Society *Brand Book* 23, no. 2 (Summer 1985): 1–7.

Raymond, Dora Neill. *Captain Lee Hall of Texas.* 1940. Norman: University of Oklahoma Press, 1982.

Robinson, Charles M., III. *The Men Who Wear the Star: The Story of the Texas Rangers.* New York: Random House, 2000.

Smith, Diane Solether. *The Armstrong Chronicle: A Ranching History.* Ed. Holland McCombs. San Antonio, Tex.: Corona Publishing Co., 1986.

State of Texas. *General Laws of the State of Texas,* 14th Leg., 2nd sess. Houston, Tex.: A. C. Gray State Printers, 1875.

———. House of Representatives. *Journal of the House of Representatives.* 14th Leg., 2nd sess. N.p., n.d.

———. Senate. *Journal of the Senate of Texas.* 14th Leg., 2nd sess. N.p., n.d.

Steele, William. *Report of the Adjutant General of the State of Texas for the Year Ending August 31, 1876.* Galveston, Tex.: Shaw & Blaylock, 1876.

Stephens, Robert W. *Texas Ranger Sketches.* [Dallas, Tex.:] Privately printed, 1972.

Taylor, T. U. "New Light on John Wesley Hardin." *Frontier Times* 2, no. 11 (August 1925): 16–19.

Thirty-Sixth Annual Meeting of the Cattle Raisers Association of Texas. Fort Worth, Texas 1912. Fort Worth: Texas Printing Co., 1912.

Thrapp, Dan L. *Encyclopedia of Frontier Biography.* Vol. 3. Spokane, Wash.: Arthur H. Clark Co., 1990.

Tise, Sammy. *Texas County Sheriffs.* Hallettsville, Tex.: Tise Genealogical Research, 1989.

Trachtman, Paul. *The Gunfighters.* Time-Life Old West Series. New York: Time-Life Books, 1974.

Tyler, Ron, ed. *The New Handbook of Texas.* 6 vols. Austin: Texas State Historical Association, 1996.

U.S. Congress. House of Representatives. *Texas Frontier Troubles.* Report no. 343, vol. 1709, Serial Set. 44th Cong., 1st sess., Washington, DC, 1876.

Utley, Robert M. *Lone Star Justice: The First Century of the Texas Rangers.* New York: Oxford University Press, 2002

Webb, Walter Prescott. *The Texas Rangers: A Century of Frontier Defense.* 1935. Austin: University of Texas Press, 1977.

———, ed. *The Handbook of Texas.* Austin: Texas State Historical Association, 1952.

Weiss, Harold J., Jr. "The Texas Rangers Revisited: Old Themes and New Viewpoints." *Southwestern Historical Quarterly* 98, no. 4 (April 1994): 620–40.

Wilkins, Frederick. *The Law Comes to Texas: The Texas Rangers, 1870–1901.* Austin, Tex.: State House Press, 1999.

Williamson, Edward C. "William Dudley Chipley: West Florida's Mr. Railroad." *Florida Historical Quarterly* 25, no. 4 (April 1947): 333–55.

Index

ISBN 978-1-58544-553-0

52000

9 781585 445530